Active Faults and Nuclear Regulation

Yasuhiro Suzuki

Active Faults and Nuclear Regulation

Background to Requirement Enforcement in Japan

Springer

Yasuhiro Suzuki
Disaster Mitigation Research Center
Nagoya University
Nagoya, Aichi, Japan

ISBN 978-981-15-0764-9 ISBN 978-981-15-0765-6 (eBook)
https://doi.org/10.1007/978-981-15-0765-6

Based on a translation from the Japanese language edition: *Genpatsu to Katsudanso* by Yasuhiro Suzuki,

This Springer imprint is published by the registered company Springer Nature Singapore Pte Ltd.
The registered company address is: 152 Beach Road, #21-01/04 Gateway East, Singapore 189721, Singapore

Preface

Why have active faults at nuclear power plant sites been overlooked in Japan? Who should evaluate the risk of active faults in the future and how should this be done? The author has collaborated with the Nuclear Regulation Authority, which was founded in the aftermath of the Fukushima nuclear disaster, in the capacity of an external expert and has investigated active faults at nuclear power plant sites and pointed out inadequacies in the previous regulatory system. This book discusses the need for change in nuclear regulations that carefully consider the latest scientific knowledge as well as the "limitations of science."

This book is a translation of *Nuclear Power Plants and Active Faults - "Unanticipated" is Unacceptable*, which I wrote and published through Iwanami Shoten in 2013 as part of the Iwanami Science library series. In the aftermath of the Fukushima nuclear disaster in 2011, the Nuclear Regulation Authority (NRA) was established in 2012, and the method of active fault assessment and regulatory requirements were examined. These requirements have been finalized and new reforms have started since 2013. In this book, the author points out the problems of assessment on active faults since the twentieth century and also discuss the direction of reform in the NRA.

Nagoya, Japan Yasuhiro Suzuki

Contents

Introduction

The day after the Great East Japan Earthquake in 2011, the Fukushima Daiichi Nuclear Power Plant operated by the Tokyo Electric Power Company (TEPCO) underwent a massive explosion. The cause was the M9 interplate earthquake at the Japan Trench and the subsequent associated giant tsunami. Preventive measures against such severe hazards had not been implemented at the power plant site because similar phenomena had not occurred in the previous 500–1000 years [1]. However, subsequently, concern has been raised regarding the safety of nuclear power plants in the event of large earthquakes and tsunamis, which has highlighted the requirement for fundamental change regarding appropriate safety measures.

Since the 2011 Great East Japan Earthquake, many other earthquakes have occurred in inland regions of the country, which have led to increased attention being focused on active faults. Earthquakes occur because of fault displacement. Faults predicted to reactivate in the future are called active faults. When faults slip on land, they cause intense local shaking, which can rupture the ground and destroy buildings and other structures. However, such situations are difficult to manage because the interval between events is generally too long for serious consideration of the implementation of possible countermeasures intended to prevent such terrible catastrophes.

General concern over where the next major earthquake might occur was not the only reason for the increased attention on active faults. Soon after the occurrence of the 2011 Great East Japan Earthquake, researchers began questioning whether the operators of nuclear power plants might be disregarding active faults that exist within their sites. In the summer of 2012, further fear among the public was prompted by frequent media reports claiming that the existence of active faults beneath some nuclear power plants might have been overlooked. In Japan, regulations require meticulous scrutiny for active faults when constructing nuclear power plants. However, if active faults are being overlooked, it is important to determine the underlying reasons, propose how the issue might be resolved, and establish who should assess active faults to ensure future safety.

In an attempt to incorporate lessons learned from the Fukushima Daiichi nuclear disaster, the Nuclear Regulation Authority (NRA) was established in the fall of

2012 as an external bureau of the Ministry of the Environment. The NRA is a highly independent organization detached from the Ministry of Economy, Trade, and Industry, which promotes nuclear energy. The NRA began investigating the problem of active faults within nuclear power plant sites, and it started taking steps to establish safety regulations.

Prior to the establishment of the NRA, regulations regarding seismic safety design were published by the former Nuclear Safety Commission, which was attached to the Cabinet Office. These regulations were bylaws of government authorities that did not possess external binding power. However, the regulation criteria "New Regulatory Requirements for Light-Water Nuclear Power Plants" (hereafter, "New Regulatory Requirements"), specified by the NRA in June 2013, are of ministerial ordinance level and they do have legal force. As part of this regulatory framework, a detailed review guide or "Detailed Note on Requirements" was also stipulated as part of the internal rules of the NRA. Thus, a strict review process ensued after July 2013.

Since 2012, the author has been part of a team discussing the content of the regulatory requirements regarding earthquakes and tsunamis. This team has reexamined the former regulations, and it has proposed revisions intended to increase robustness. In the process, the team was criticized for instituting new and excessively strict requirements regarding active fault criteria. Others predicted the new criteria would lead to an increased number of faults being classified as active faults. However, the new requirements and criteria were simply a continuation of the former rules; the primary focus of our team was their rigorous enforcement. Nevertheless, the repercussions were vast, which suggested that the former regulations had not been stringently enforced.

A prominent former regulation was the "Regulatory Guide for Reviewing Seismic Design" [2], which was revised in 2006 (hereafter, "Revised Regulatory Guide 2006"). This includes clear statements designed to prevent oversight in relation to active faults. However, this regulation was not strictly enforced in the review process because of concerns regarding the impact on existing nuclear power plants. In addition, the fundamental aspect of who is responsible for evaluating nuclear power plant safety and for guaranteeing it for the public has remained ambiguous for a long period, which has likely caused further underestimation of active faults.

This publication illustrates the discussions that took place at the NRA concerning active faults, examines past issues in relation to active faults, and considers how these problems should be resolved. The public cannot be assured of the safety of nuclear power plants without resolving past issues. There are inherent limitations regarding the extent to which we can understand active faults. The principal difficulty lies in how to make strict and reasonable judgments given current scientific limitations. The key question is whether we can overcome this issue and correct the course of safety regulations with regard to nuclear power plants.

References

1. Suzuki Y (2012) Social demand level of hazard estimation and the responsibility. Gakujutsu-no Doko 17(8):20–24 (in Japanese)
2. Nuclear Safety Commission (2006) Regulatory guide for reviewing seismic design of nuclear power plant facilities. https://www.nrc.gov/docs/ML0803/ML080310851.pdf

Chapter 1
The Active Fault Problem in Relation to Nuclear Power Plants: Current Issues

Abstract The explosion of the Fukushima Daiichi Nuclear Power Plant in 2011 due to the Great East Japan Earthquake led to the initiation of a comprehensive inspection of the earthquake resistance of every nuclear power plant in Japan. In addition to protection against tsunamis induced by interplate earthquakes, the inspections covered resistance against extremely strong vibrations and surface deformations caused by earthquakes. Reconsideration of the safety regulations for Japan's nuclear power plants actually began with the 1995 Kobe earthquake, and in 2006 the safety review guidelines were strengthened. However, in a reexamination ("back-check") after 2006, strict examination based on the new guidelines was not performed. The sensitive location of the reactor (the reactor is located directly above the active fault) was also overlooked. The situation has been described as a "regulatory capture" by the Fukushima Nuclear Accident Independent Investigation Commission of the National Diet. In addressing the reasons behind the lack of strict screening, the ambiguity and arbitrariness of the definition of active faults seem to be the fundamental problems.

Keywords Active faults · Regulatory capture · Just-on-fault problem · Risk awareness

1.1 The Need for Inspections on Active Faults

The Fukushima Daiichi nuclear disaster is thought to have occurred not because of earthquake-related strong ground motion but because of the loss of emergency power following the associated tsunami. Earthquake measures should be reexamined holistically because the occurrence of such a huge earthquake was "unanticipated." In fact, some theories consider that part of the damage leading to the disaster was due to ground motion [1].

Y. Suzuki, *Active Faults and Nuclear Regulation*,
https://doi.org/10.1007/978-981-15-0765-6_1

Giant tsunamis such as the one associated with the 2011 Great East Japan Earthquake have occurred previously. In the Jogan earthquake of 869, a tsunami swept up the coast from Iwate Prefecture to Fukushima Prefecture. The wave run-up in this event extended 3 km inland in the region of the Sendai Plains. The magnitude of the associated earthquake was estimated to be at least M8.4 [2, 3]. In addition, a recent study on active sea-bottom faults found a giant fault near the trench associated with the 2011 earthquake that showed topographic evidence of reactivation. This suggests that similar earthquakes might have occurred repeatedly in the past [4, 5]. Thus, these findings imply that the huge 2011 earthquake was not necessarily a completely unpredictable phenomenon.

Caution regarding giant tsunamis was more relaxed in regions south of Miyagi Prefecture compared with Iwate Prefecture. This was because giant tsunamis had not occurred in regions south of Miyagi Prefecture since the Meiji Era at the end of the nineteenth century. Considering that a giant tsunami such as the one associated with the Great East Japan Earthquake occurs only once every 500–1000 years, it might be difficult to accuse the government of a lack of consideration for such events within its general disaster measures prior to 2011. However, this was clearly not appropriate regarding seismic design for important facilities such as nuclear power plants, even before 2011. Huge, low-frequency disasters that occur once in a millennium tend to be ignored when considering the level of rigor for disaster measures, principally in the belief that such disasters will not occur or in an attempt to minimize costs. This stance should not be tolerated. TEPCO and the former Nuclear and Industrial Safety Agency ("Safety Agency"; a specialized agency established in 2001 as part of the Agency for Natural Resources and Energy, which is an external bureau of the Ministry of Economy, Trade, and Industry) both forecasted the possibility of a tsunami larger than 15 m occurring at the Fukushima Daiichi Nuclear Power Plant. However, this was disregarded as a primary target of the disaster prevention measures. Hence, in this case, "unanticipatcd" actually meant "willfully disregarded," as was verified by various media reports following the earthquake [6].

Active faults are representative examples of phenomena that cause huge, low-frequency disasters. They have become a topic of focus in reconsideration of earthquake measures in relation to nuclear power plants. Following the occurrence of the Kobe Earthquake in 1995, triggered by active faults beneath the metropolis of Kobe (i.e., the Rokko-Awaji fault zone), comprehensive inspection of nuclear power plants began amid concern regarding their resistance to earthquakes (Fig. 1.1).

Prior to the issuance of Revised Regulatory Guide 2006 by the former Nuclear Safety Commission, the Regulatory Guide enacted in 1978 was in use. In addition, the publication of "Guidelines for Reviewing Seismic Safety Concerning Geological Faults" was produced in 2008. (In 2010, this was merged into the "Guidance for Reviewing Seismic Safety of Nuclear Power Reactor Facilities" (hereafter, "Review Guidelines 2010").) While discussing the implementation of these rules, we realized there were many issues concerning the identification of active faults and that considerable numbers of active faults had been overlooked.

The event that catalyzed these developments occurred in 2006, when Dr. Takashi Nakata (Professor Emeritus, Hiroshima University) discovered evidence of an

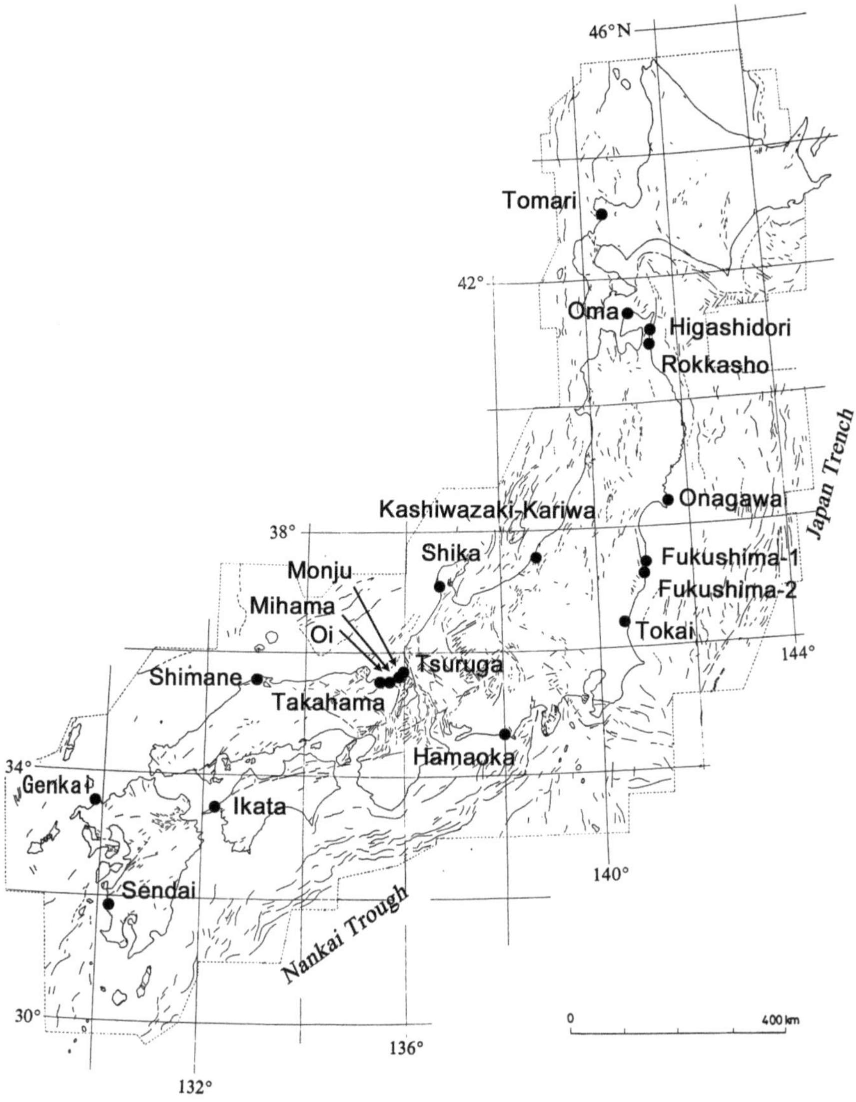

Fig. 1.1 Active faults and nuclear power plants in Japan. Translated from Suzuki [7] @Yasuhiro Suzuki 2013

active fault that the Chugoku Electric Power Company had deemed inactive. Subsequently, in 2007, the Noto Hanto Earthquake and the Niigata-ken Chuetsu-oki Earthquake struck, the origins of which were near the Shika Nuclear Power Plant operated by the Hokuriku Electric Power Company and TEPCO's Kashiwazaki-Kariwa Nuclear Power Plant, respectively. Thus, when reviewing the seismic design to verify the safety of these nuclear power plants, a detailed survey was conducted on the presence of active sea-bottom faults that could potentially cause earthquakes (active sea-bottom fault survey). Nevertheless, the presence of significant active faults that caused earthquakes had been overlooked [8].

After the 2011 Great East Japan Earthquake, an M7.0 earthquake occurred in Hamadori, Fukushima, on April 11. This earthquake, which caused evident surface ruptures, originated on the Itozawa Fault and on the Yunodake and Shionohira faults (Fig. 1.2) [9]. Although the Shionohira Fault was previously unknown, the Itozawa and Yunodake faults were discovered by active fault researchers in 1991 (Fig. 1.3). However, although TEPCO classified the Itozawa Fault as active during the seismic design review of the Fukushima Daiichi Nuclear Power Plant, they classified the Yunodake Fault as inactive. The former Safety Agency and former Nuclear Safety Commission also regarded those interpretations as appropriate. The main reason was that the Yunodake Fault was considered a normal fault that would not reactivate (and that the fracture zone was fixed). This was based on the understanding that normal faults move under tensile stress and that normal faults in most of the Japanese Archipelago should not move because of the compressive stress field. Similar judgments were made in relation to several other nuclear power plants. However, these judgments lost validity after a normal fault in Hamadori (Fukushima) moved and caused an earthquake.

Fig. 1.2 Obvious surface rupture along Yunodake Fault. Photo is reprinted by courtesy of Dr. Hiroyuki Tsutsumi

The repercussions of finding an active fault at a nuclear power plant site are considerable. In August 2012, the former Safety Agency instructed comprehensive inspections be undertaken on fracture zones (areas in which rocks become crushed because of fault activity) [10]. As the meaning of the term is similar to that of faults, the term faults is used hereafter. Specifically, existing documents suggested that nuclear power plants at Tsuruga (Japan Atomic Power Company), Shika (Hokuriku Electric Power Company), Higashidori (Tohoku Electric Power Company), Oi (Kansai Electric Power Company: KEPCO), Mihama (KEPCO), and the Fast Breeder Reactor Monju (Japan Atomic Energy Agency) likely had issues. Consequently, the NRA undertook detailed fault surveys at these power plants.

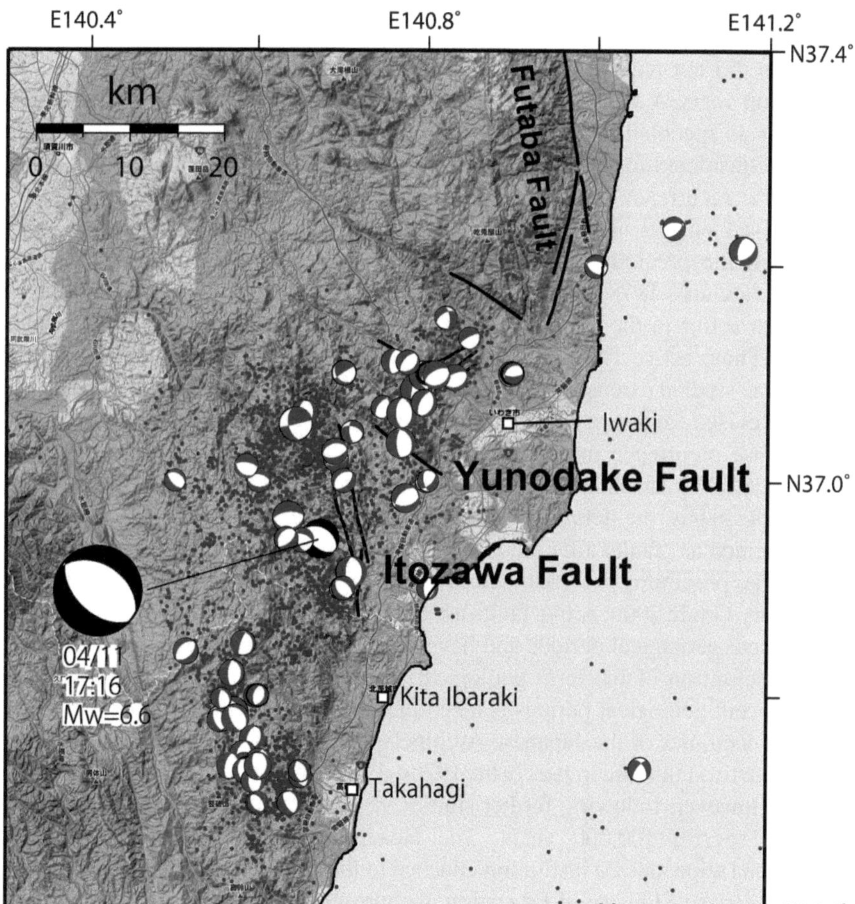

Fig. 1.3 Fukushima-ken Hamadori Earthquake of 11 April, 2011, and its causative active faults. Circle shows their seismic mechanism. Translated from Tsutsumi and Toda [9] with the permission of Geological Society of Japan

1.2 How Are Active Faults Defined?

The definition of an active fault is an issue when examining whether active faults are identified correctly and taken into account in seismic design. Appropriate judgments cannot be made if the definition is vague or arbitrary; this is where the biggest cause of confusion lies. The question of how an active fault should be defined to ensure nuclear power plant safety is not complicated, and the wording has been defined in the former rules, including the Revised Regulatory Guide 2006 and Review Guidelines 2010. However, the desire to classify faults as not active whenever possible appears to influence the interpretation of the definition and to invite confusion.

In the first version of the regulatory guide ("Regulatory Guide 1978") [11], active faults are defined as "faults that have shown activity in the Quaternary Period (in the past 1.8 million years) and have potential for activity in the future." However, this definition did not require that evaluation on activity should encompass the entire 1.8 million years. A separate definition for "active faults to be examined in seismic design" was specified from the perspective of the period of geological time that must be considered for effective identification of faults that might reactivate in the future. The recurrence interval of active faults tends to be around 50,000 years at most. Based on this understanding, the general rule that faults that have shown activity in the previous 50,000 years should be evaluated was included in the Regulatory Guide 1978. This was based on the idea that faults necessarily had to have been active in the previous 50,000 years if they had any potential for future activity. Thus, active faults would not be overlooked as long as this geological period was studied extensively.

The Revised Regulatory Guide 2006 expanded the geological period to be studied because recurrence intervals can span periods longer than 50,000 years, and because 50,000-year-old geological strata and formations are not necessarily clearly identifiable. Thus, the definition "active faults to be examined for seismic design" was redefined as "faults after the late Pleistocene (120,000–130,000 years ago) for which the possibility of past activity cannot be eliminated." In the Revised Regulatory Guide 2006, active faults are defined as "faults that have repeated activity in recent geological periods and have potential for activity in the future." Here, precise definition of the term geological periods is not specified. Generally, the phrase recent geological periods is taken to mean "past periods in which the stress field and tectonics of the Japanese Archipelago are thought analogous to the present." For active fault researchers in the 1970s–1980s, this meant the entire Quaternary Period. However, following further studies, researchers came to regard this as the previous several 100,000 years. The Headquarters for Earthquake Research Promotion (a specialized institution attached to the Prime Minister's Office and currently under the Ministry of Education, Culture, Sports, Science and Technology), which was established to implement knowledge learned from the Kobe Earthquake and to promote earthquake research nationwide, currently sets the threshold at about 400,000 years [12].

To assess seismic safety in nuclear power plants, the definition of "active faults to be examined for seismic design"[1] is of crucial importance. Although interpretation of the applicable geological period has changed, as described above, the most significant amendment in the Revised Regulatory Guide 2006 was that the criteria came to include faults for which "the possibility of activity cannot be eliminated." Such a statement was not present in the older version of the guidelines (i.e., Regulatory Guide 1973); thus, faults could be classified as inactive if definitive evidence of activity over the previous 50,000 years could not be found. Conversely, Revised Regulatory Guide 2006 requires faults be classified as active unless activity in the past 120,000–130,000 years can be discounted. This has garnered attention as a decision-making policy that emphasizes safety. The Review Board (the seismic guideline sectional committee established under the Nuclear Safety Commission's Nuclear Safety Standards Subcommittee in 2001) also confirmed the intended meaning of this statement numerous times and it left clear records in the minutes of its meetings. Put simply, the decision of the Review Board was to have items in the "gray zone" evaluated as "black," i.e., to classify capable faults with uncertain activity as active faults.

Unfortunately, the intentions behind Revised Regulatory Guide 2006 were not acted upon. Some evaluations interpreted active faults in a narrow sense, e.g., considering normal faults to be inactive. Some former committee members who reviewed nuclear power plant safety have argued that only a seismogenic fault (a fault that causes earthquakes) can be classified as an active fault. Indeed, in academia, there are various differences in the interpretation of what constitutes an active fault. Those whose research interests are focused on large-scale geological change or seismic prediction tend to regard only relatively large seismogenic faults as active faults. Nuclear regulations, however, do not include wording to that effect.

It can be argued that faults that appear only in shallow areas should not be classified as active faults because seismogenic faults should extend to deeper areas. However, this is problematic, because if such a fault were discounted as an active fault and a facility constructed over it, seismic-related ground movement could result in serious consequences for any buildings. Thus, one cannot disregard faults that do not extend to deeper areas. Researchers have particular ways of understanding a term or have individual wording preferences; thus, the precise definition of what constitutes an active fault is inconsistent. It is true that previous regulatory reviews did not clarify this point, which promoted certain ambiguity in its implementation. Clearly, the lack of explicit rules that could be strictly enforced was a fundamental problem.

The definition of what constitutes an active fault might have become increasingly specific after Regulation Guide 1978 was introduced, when more was known about active faults. License applications written before the guidelines were established included serious debate over whether faults were active or inactive, regardless of

[1] The wording "active faults to be examined for seismic design" was replaced by "faults that might exhibit activity in the future" because of the establishment of the updated "Regulatory Requirements 2013" by the NRA, as detailed in Chap. 5.

size or character [13]. Revised Regulatory Guide 2006 should have been a significant turning point that brought the opportunity to reevaluate active faults. It would be extremely problematic if reviews were nevertheless carried out in the same manner as before in re-inspections of all nuclear power plants (i.e., the so-called back-check). If reviews took into account the influence on existing nuclear power plants, this could well be a case of regulatory capture (in which the regulator is practically governed by the licensee), as was highlighted in the official report of the Fukushima Nuclear Accident Independent Investigation Commission by the National Diet of Japan. (https://www.nirs.org/wp-content/uploads/fukushima/naiic_report.pdf).

1.3 The Problem of Active Faults Under Nuclear Power Plants

Although it is widely believed that active faults slip rarely, an unlikely but possible disaster at a nuclear power plant could cause irreversible damage. It might happen only once in a millennium, but the fault will surely slip at some point, which is why nuclear power plants should not be built on top of active faults.

A brochure intended to highlight the safety of nuclear power plants for the public clearly stated in "Seismic Measure No. 1" that nuclear power plants would not be built on top of active faults [14]. However, official regulations remained ambiguous for a long period. The "Regulatory Guide for Reviewing Nuclear Reactor Siting Evaluation and Application Criteria" (introduced in 1964 by the Nuclear Safety Commission), which outlines criteria for assessing suitability in nuclear power plant sites, states that as a fundamental rule, sites should "not have had past phenomena that would cause major disasters and have no possibility of such phenomena happening in the future." This could be interpreted as a directive prohibiting the construction of nuclear power plants over active faults. However, its enforcement is problematic because active fault inspections are undertaken for the first time in the seismic review, which follows the site review.

This issue was debated extensively during June–October 2010 when the revising of guidelines for the safety review of active faults [15] was being discussed. I also took part in the discussion. Details of the discussion are available on the website of the former Nuclear Safety Commission (currently of the NRA) in a section titled "Views on the revision of the seismic safety review guidelines concerning geological faults" [16]. The records of these meetings are significant because similar discussions were held again by the NRA after 2012. A summary of the meeting proceedings is outlined in the following with additional explanatory comments.

During the 16th Guideline Review Committee meeting (June 3, 2010), committee member Dr. Takashi Nakata proposed the addition of the phrase "Active faults or faults in which the possibility of activity cannot be eliminated must be absent from the foundation ground of nuclear reactors." In response, some members expressed opposition, stating, "This phrase should not be added because the important thing is the safety of the facility itself in terms of engineering (author comment: e.g., a reactor can be built over an active fault as long as it will not collapse)." The issue was then carried over to the next meeting.

During the 17th and 18th Guideline Review Committee meetings (July 9 and August 4, 2010, respectively), the secretariat expressed the position that "When assessing the supporting ability of foundation grounds in which buildings or structures are built on, strong shaking occurring from fault activity, as well as fault displacement associated with earthquakes, should be considered." Thus, reconsideration of the definition of active faults began. Large active faults cause earthquakes through independent movement. Small active faults do not cause earthquakes on their own; however, they do slip in association with other faults when earthquakes occur. To highlight how not only the former but also the latter is a crucial point to consider, Dr. Nakata and I stated that "Not only seismogenic faults but also faults that could cause slip (capable faults), which the IAEA's guidelines (Safety Guide) mention, should be considered." We also asserted that "In addition to faults that might cause earthquakes (causative faults), faults that are structurally relevant (secondary faults) should clearly also be considered active." One committee member disagreed, claiming that "It is clear from past developments that the definition of active faults in the seismic design regulatory guide only includes potentially seismogenic faults; not all capable faults have to be considered active." He also added, "We should keep in mind that IAEA regulations take into account various situations across numerous countries." Did this mean that in earthquake-ridden Japan, it is impossible to deal with every fault with care?

Some members expressed the opinion that "If this prohibition clause is documented, the various possible interpretations of what constitutes an 'active fault' or 'outcrop' would be an issue. Therefore, it is better not to specify this and instead allow each to be reviewed on an individual basis during the safety review phase." In addition, others said, "We should set restrictions to prevent broad interpretation and instead focus on the point that sites that clearly compromise safety should be avoided." As these statements illustrate, a number of members were against uniform prohibition, because they believed that depending on the level of displacement and deformation, facilities could be engineered to be safe, even with active faults immediately beneath. However, we asserted that "Evaluating whether a facility can maintain safety is a challenge in itself, since scientifically predicting the degree of displacement and deformation is difficult." We further explained that "We have already declared to the public that construction will not take place over active faults. We should highlight this clause in the guidelines because transferring new facilities to alternative sites is an option." Finally, chief examiner Dr. Kojiro Irikura (Professor Emeritus, Kyoto University) concluded that there is meaning in clarifying the positioning of this issue for the public and the meeting ended with the declaration that "Since we all share the intention to secure safety, we shall reach a final conclusion through consideration of each opinion."

During the 19th and 20th Guideline Review Committee meetings (September 8 and October 7, 2010, respectively), the definition of "active faults to be examined for seismic design" was finalized as "faults that might cause earthquakes (causative faults), as well as structurally relevant faults (secondary faults)." With regard to the latter, the conclusion was that "appropriate evaluations must be undertaken if the possibility of displacement related to causative fault activity cannot be eliminated." Through these discussions, the final Review Guidelines specified that "If an outcrop of an active fault that should be examined for seismic design is identified, clauses regulating the 'assessment of a foundation's supporting ability' do not apply, as the construction of buildings or establishments with Importance Category S class above such faults are unanticipated." In other words, this means that significant establishments would not be built on top of active faults in the first place, and in the event that turns out to be the case, the situation would not be considered acceptable, even if the facility should be safe when the fault moves.

After the Great East Japan Earthquake, the problem of underlying active faults gathered attention. As a fault that TEPCO considered inactive underwent significant displacement in the Fukushima-ken Hamadori Earthquake, which occurred a month after the 2011 Great East Japan Earthquake, the former Safety Agency carried out nationwide re-inspections of active faults near nuclear power plants. When they reexamined existing documents, they highlighted that potential active faults might have been overlooked in some cases. Inspectors fired sharp criticisms, saying, "This is clearly a typical active fault" and "How in the world did this plant pass the review? I'm speechless." These comments were picked up by the media and they led to the focus of public attention on the problem of underlying active faults [17].

That there were no clear regulations prohibiting the construction of important facilities over active faults and that the revised guidelines only included the expression "unanticipated" were also seen as problematic. The public realized that the word unanticipated had clearly been used to avoid responsibility and the wish was expressed that this should not happen again in the future. As a continuation of such discussion, the NRA is now developing appropriate criteria. The ongoing process is illustrated in further detail in Chap. 4.

1.4 Why Are Active Faults Dangerous? The Extent of Damage and Prevention Measures

Active faults in Japan are thought to have recurrence intervals of roughly a thousand to several tens of thousands of years. Some, such as the Tanna and Yoro faults located in central Japan, might have repeated activity over periods of less than a thousand years; however, they have shown activity only once or twice in recorded history [18]. Fault activity is infrequent and active faults do not slip gradually on a regular basis. Thus, it is extremely rare for an active fault to cause any disruption to daily life. However, once an active fault does move, it can cause intensity 7 (maximum scale in Japan) tremors locally, rupturing the ground surface and at times slipping several meters. When discussing damage associated with an active fault, we must consider shaking and rupturing separately.

With regard to shaking, the 1995 Kobe Earthquake left a "severely damaged zone" along the active fault. Active faults that became the origin of this event were the Nojima Fault in Awaji Island and the Rokko Fault in the Kobe area. Intense shaking occurred within a 1-km zone along these faults, which caused many buildings and civil engineering structures to suffer significant damage. Actually the Rokko Fault consists of several faults. Since obvious surface earthquake faults did not appear, it remains unclear which fault moved. After the earthquake, Dr. Mitsuhisa Watanabe and I discovered an active fault underneath the severely damaged zone and we suggested that its activity might have played a part in the destruction [19]. The widely accepted "Basin Edge Effect Theory" [20, 21] explains how the unique geological structures under the plains in Kobe can amplify shaking. However, this theory assumes the absence of preexisting faults underneath the severely damaged zone. Therefore, this does not refute the hypothesis that fault activity also influenced shaking, in addition to the amplification effect of the geological structures.

The development of areas of serious damage was also evident in the 1945 Mikawa Earthquake and the 1948 Fukui Earthquake in Japan. More recently, the same can be said of the 2004 Niigata-ken Chuetsu Earthquake (and of the 2016 Kumamoto Earthquake). The Headquarters for Earthquake Research Promotion has made public a map showing the distribution of predicted shaking intensity when major active faults move [22]. In any case, maximum shaking intensity of such events is estimated at around intensity 7. At this level of intensity, simple adherence to the Building Standards Act would not prevent severe damage to buildings. Existing buildings that do not meet the requirements of the Building Standards Act would very likely collapse.

Serious damage can also occur because of rupturing of the ground along the line where active fault planes merge with the ground surface. All buildings built on the fault line collapsed in Aichi Prefecture during the 1945 Mikawa Earthquake and on Awaji Island during the 1995 Kobe Earthquake. In addition, in the 1999 Chi-Chi Earthquake in Taiwan, a school built over the fault line was destroyed, and a number of bridge piers and dams collapsed (Fig. 1.4). Rupturing above faults cannot be prevented, which means building collapse is inevitable regardless of the strength and reinforcement of a building and its foundations.

In the state of California (USA) and in New Zealand, the use of land on fault lines is highly regulated, and the construction of housing establishments beyond a certain size is prohibited. Similar regulations were enforced in Taiwan after the earthquake in 1999. Curiously, the Japanese government has not taken any such measures, even though the country has experienced considerable damage from many previous earthquakes. One problem is that the levels of damage caused by shaking and by rupturing were considered in combination when the need to regulate land use was discussed. In response to opinions that affirmed the need for land use regulation [23], some said, "We cannot decide how far buildings have to be from fault lines to avoid damage." Others claimed that "Even if buildings are not constructed along fault lines, damage will be serious if the foundation ground is weak; therefore, it is not reasonable to restrict construction only along fault lines." These objections, however, appear to have been expressed because of social concerns, e.g.,

Fig. 1.4 Damages due to the 1999 Chi-Chi Earthquake in Taiwan. (**a**) Collapsed school building on the surface fault. (**b**, **c**) Damages of bridges on the surface fault. Reprinted from Suzuki [18]@ Yasuhiro Suzuki 2001

such as the potential for a fall in land prices, rather from because of doubts about the effectiveness of lessening the extent of damage. Some clearly misunderstood the facts, saying, "Measures are necessary, but they cannot be enacted since the locations of active fault lines cannot be determined beforehand."

Since 1995, the Geospatial Information Authority of Japan has been developing a detailed, nationwide map of active faults so that anyone can ascertain the locations of active faults [24]. Consequently, responsibility can no longer be avoided by playing the "unanticipated" card in the event that an active fault activates. Privately owned buildings aside, management responsibility already exists for owners of public facilities such as schools and hospitals, and there have been attempts in the past to relocate facilities such as schools [25]. In 2013, Tokushima Prefecture enacted a regulation that demanded construction of public facilities be prohibited within a 40-m designated area around an active fault, where the line of the active fault is clear [26]. Even if the probability of faulting and subsequent damage is low, the predictability remains considerably high. Therefore, if an event does occur, any omission in management responsibility might be questioned. Although some say that "restricting only pre-discovered active faults is a problem since not all of them can be discovered in advance," this depends on how one understands the limitations of science. There is ample opportunity for averting risk, such as relocation; thus, it is unreasonable to ignore active faults that have already been discovered.

References

1. National Diet of Japan Fukushima Nuclear Accident Independent Investigation Commission (2013) Report. http://warp.da.ndl.go.jp/info:ndljp/pid/3514027/naiic.go.jp/pdf/naiic_honpen.pdf (in Japanese) https://www.nirs.org/wp-content/uploads/fukushima/naiic_report.pdf (in English)
2. Satake K, Namegaya Y, Yamaki S (2008) Numerical simulation of the AD 869 Jogan tsunami in Ishinomaki and Sendai plains. Annual Report on Active Fault and Paleoearthquake Researches 8:71–89. (in Japanese)
3. Namegaya Y, Satake K (2014) Reexamination of the A.D. 869 Jogan earthquake size from tsunami deposit distribution, simulated flow depth, and velocity. Geophys Res Lett 41:2297–2303
4. Nakata T, Goto H, Watanabe M, Suzuki Y, Nishizawa A, Izumi N, Ito K (2011) Active faults along the Japan trench and large earthquakes, Japan Geoscience Union Meeting, MIS036-P189
5. Nakata T, Goto H, Watanabe M, Suzuki Y, Nishizawa A, Izumi N, Horiuchi D, Kido Y (2012) Active faults along Japan trench and source faults of large earthquakes, Proceedings of the international symposium on engineering lessons learned from the 2011 Great East Japan Earthquake, 254–262
6. Shimazaki K (2011) Huge tsunami that was anticipated but left out of consideration. Kagaku 81:1002–1006. (in Japanese)
7. Suzuki Y (2013) Nuclear power plants and active faults – "unanticipated" is unacceptable. Iwanami Shoten, Tokyo (in Japanese)
8. Suzuki Y, Nakata T, Watanabe M (2008) Essential problem of active fault assessment in safety regulation of nuclear power plants in Japan. Kagaku 78:97–102. (in Japanese)
9. Tsutsumi H, Toda S (2012) Surface rupture associated with the April 11, 2011 Fukushima-ken Hamadori earthquake and Paleoseismic history of the causative fault. J Geol Soc Jpn 118:559–570. (in Japanese)
10. Nuclear and Industrial Safety Agency (2012) Press Release. http://www.meti.go.jp/press/2012/08/20120829006/20120829006.html (in Japanese)
11. Nuclear Commission (1978) Regulatory guide for reviewing seismic design of nuclear power plant facilities (in Japanese)
12. Headquarters for Earthquake Research Promotion (2010) Report on methods of evaluating active fault, preliminary version (in Japanese)
13. Japan Atomic Power Company (1965) License application for Unit 1 of Tsuruga Nuclear Power Plant. (in Japanese)
14. Agency for Natural Resources and Energy (1995) Brochure "Seismic safety of nuclear power plants" Published by Nuclear Power Engineering Corporation. (in Japanese)
15. Nuclear Safety Commission (2008) Guidelines for reviewing seismic safety concerning geological faults. (in Japanese)
16. Nuclear Safety Commission, Subcommittee for Review Guideline (2010) Views on the revision of the seismic safety review guidelines concerning geological faults. http://www.nsr.go.jp/archive/nsc/senmon/shidai/chishitsu/chishitsu023/ksiryo1.pdf (in Japanese)
17. Mainichi Shimbun (2012) Shika Nuclear Power Plant, Morning paper on 18th July, 2012. (in Japanese)
18. Suzuki Y (2001) Prevention against earthquakes by active faults. Kokon-shoin. (in Japanese)
19. Suzuki Y, Watanabe M, Azuma T, Okada A (1996) Active tectonics of the Rokko-Aaji area, and the earthquake fault of the southern Hyogo prefectural earthquake of 1995: tectonic geomorphological studies and problems. Geographical Review of Japan, Series A 69:469–482. (in Japanese)
20. Irikura K (1996) Strong ground motion causes "severely damaged zone". Kagaku 66:86–92. (in Japanese)
21. Pitarka A, Irikura K, Iwata T, Kagawa T (1996) Basin structure effects in the Kobe area inferred from the modeling of ground motions from two aftershocks of the January 17, 1995, Hyogo-ken Nanbu earthquake. Journal of Physics of the Earth 44:563–576

22. Headquarters for Earthquake Research Promotion. Evaluation of strong ground motion. https://www.jishin.go.jp/evaluation/strong_motion/ (in Japanese)
23. Nakata T (1990) Alquist-Pliolo special studies zones act of California and prevention of earthquake hazards. J Geogr 99:289–298. (in Japanese)
24. Geospatial Information Authority of Japan. Active Fault map. http://www.gsi.go.jp/ENGLISH/page_e30084.html.
25. Suzuki Y (2009) Anti-seismic countermeasures of school buildings for active faults. Kagaku 79:175–178. (in Japanese)
26. Tokushima Prefecture (2015) Designation of special research area for active fault. https://anshin.pref.tokushima.jp/docs/2013082700025/ (in Japanese)

Chapter 2
Reasons Why the Threat of Active Faults Has Been Neglected in Japan

Abstract The characteristics and scale of active faults are important for predicting future earthquakes. The positional relationship between an active fault and a nuclear site presents information that predicts not only the magnitude of earthquake but also the ground disaster at the nuclear site. Overlooking the active fault trace or underestimating its length may directly lead to a nuclear accident. However, active faults have been underestimated in Japan's nuclear seismic assessment; the reasons were low awareness about the risk of active faults, overconfidence in earthquake-resistant construction, and misunderstandings regarding the rationality of low-frequency disaster countermeasures. The safety review did not take into account the "limitations of science," but relied only on the information found in the survey. There were also problems with the system in which only operators conducted investigations and compiled conclusions, screening system with only certain members, and vague screening rules that allowed arbitrary interpretation.

Keywords Underestimation of active faults · Limitations of science · Low-frequency disasters · Review systems · Ambiguity of safety review rules

2.1 Problems with the Understanding and Perception of Active Faults

The threat of active faults has sometimes been taken lightly during seismic design reviews for nuclear power plants. Thus, in recent years, oversight regarding active faults has become apparent in relation to some nuclear power plants in Japan. Nevertheless, some claim this is not a significant issue and assert that there were no problems with previous reviews. This is where underestimation of the threat of active faults appears to lie; there is a problem of awareness of active faults and a problem regarding the research history.

Y. Suzuki, *Active Faults and Nuclear Regulation*,
https://doi.org/10.1007/978-981-15-0765-6_2

Awareness of the threat of active faults appeared to be low prior to the 1990s. Experts that design nuclear power plants tend to believe the plants are unlikely to ever collapse, because such important facilities have substantially higher seismic resistance than other buildings and because they are built on robust foundation grounds. Even if some active faults were overlooked, nuclear power plants are designed to withstand considerable shaking. Moreover, while active faults might only cause an earthquake once in a millennium, the operating life of a nuclear power plant is around 30–40 years. Hence, the probability of an active fault exhibiting activity while a plant was in operation is close to zero. The reasoning is, under such circumstances, making robust preparations for fault activity could be considered excessive and impractical. Thus, even if the presence of active faults was known, there has been a tendency to presume they would not move. In fact, the website of one energy company publicized that "Nuclear power plants can even withstand earthquakes that are never meant to occur." When I asked what this meant, I was surprised to learn that it included earthquakes caused by active faults. This phrase is very telling of how active faults have been perceived.

The Kobe Earthquake in 1995 also had an effect on the nuclear power industry. The main reason was the collapse of numerous structures including bridge piers of expressways because of the intense level of shaking. Another reason was that the public was shaken because of the realization that an active fault had actually caused an earthquake. The 2000 Tottori-ken Seibu Earthquake, 2004 Niigata-ken Chuetsu Earthquake, 2005 Fukuoka-ken Seiho-oki Earthquake, 2007 Noto Hanto Earthquake, 2007 Niigata-ken Chuetsu-oki Earthquake, and 2008 Iwate–Miyagi Nairiku Earthquake followed soon afterward. All these earthquakes were related to inland active faults; thus, one could no longer claim that these were "earthquakes that were never meant to occur" (Fig. 2.1). Unsurprisingly, public interest in active faults skyrocketed after 1995 (Fig. 2.2).

By July 2001, the Regulatory Guide Subcommittee was established within the former Nuclear Safety Committee. After a long discussion, the Revised Regulatory Guide was produced in 2006. In June 2006, Dr. Takashi Nakata discovered an active fault following trench surveys (a survey method in which geological displacement is observed in a narrow, long trench dug across the fault) at a site where the Chugoku Electric Power Company had denied the presence of active faults. This highlighted issues surrounding the surveying of active faults. The work of Dr. Nakata had a considerable impact. The conclusion was that insufficient implementation of the so-called tectonic geomorphological methods, in which active faults are identified through tectonic landforms (as performed by Dr. Nakata and colleagues), had invited oversight of active faults. This led to a full-fledged implementation of tectonic geomorphological surveys in the "Revised Regulatory Guide 2006," the details of which are elaborated in Chap. 3.

Another reason why active faults had been overlooked concerns historical issues regarding research on active faults. The "Atomic Energy Basic Act" and the Regulatory Guide for Reviewing Nuclear Reactor Siting Evaluation and Application Criteria were enacted in 1955 and 1964, respectively. Nuclear power plant development accelerated rapidly after the 1950s and it peaked during the 1960s–1970s (the

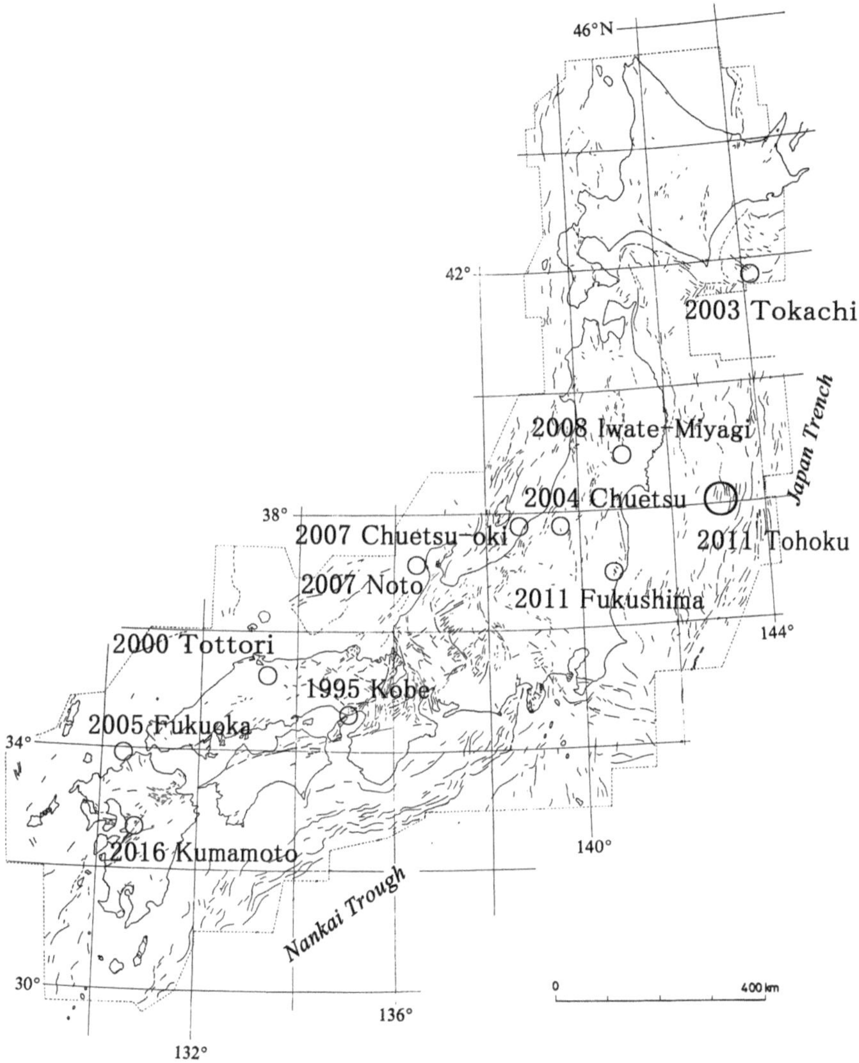

Fig. 2.1 Active faults and major earthquakes after 1995. Translated from Suzuki [1]@Yasuhiro Suzuki 2013

years of service commencement are indicated below: Tokai-mura 1966, Tsuruga 1970, Mihama 1970, Fukushima 1971, Shimane 1974, Takahama 1974, Hamaoka 1976, Ikata 1977, Oi 1979, Kashiwazaki-Kariwa 1985, and Shika 1993). In contrast, although research on active faults had been undertaken in Japan since the late Meiji Period, research that determined activity timelines and displacement amounts flourished only after the late 1960s or 1970s. Thus, active fault research had not progressed sufficiently at the time when the development of nuclear power plants had begun.

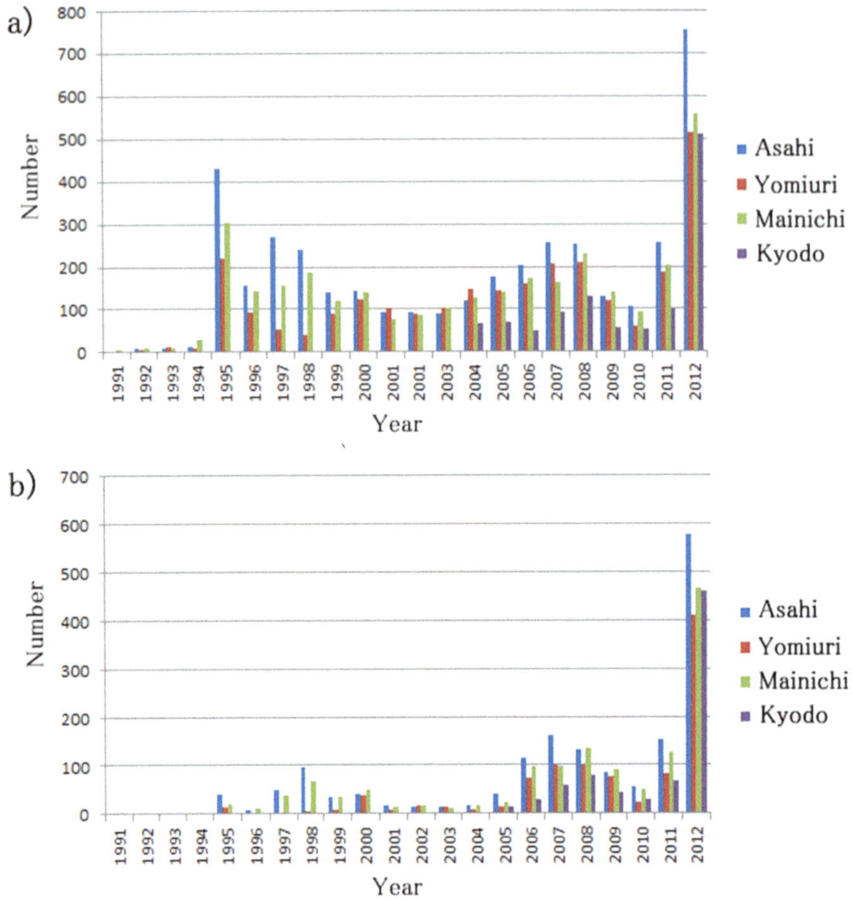

Fig. 2.2 Number of newspaper reports (1991–2012). (**a**) Reports including "active fault." (**b**) Those including both "active fault" and "nuclear power." Translated from Suzuki [1]@Yasuhiro Suzuki 2013

The course of nuclear power plant development is difficult to correct once it starts. For instance, when building Tsuruga Unit 1, the Urazoko Fault, which passes through the site, was considered inactive. This was not reevaluated when Unit 2 was added. In 2004, applications for further expansion with Units 3 and 4 were filed under the same assumption. The submitted documents included information that strongly suggested the presence of active faults. In the review of the former Safety Agency, this was considered highly problematic and a trench survey was requested. As a result, an obvious active fault was discovered. The plant operators finally admitted in 2008 that the fault was active. Their judgment was likely based on the idea that issues might arise regarding the existing reactors if the presence of the active faults was admitted when expanding the facility. From a safety security point of view, however, this should not be tolerated.

In the 1970s, Dr. Tokihiko Matsuda (Professor Emeritus, University of Tokyo) developed a relative equation between fault length and earthquake magnitude. This was utilized extensively in the seismic design of nuclear power plants [2]. In 1980, researchers published through the Tokyo University Press "Active faults in Japan" [3, 4], which was a map of the nationwide distribution of active faults with an indication of the degree of certainty expressed from highest to lowest as levels I, II, and III. However, this document was not received well, and the Nuclear Civil Engineering Committee of the Japan Society of Civil Engineers expressed concern regarding the validity of the tectonic geomorphological methods that were instrumental to the development of the map. In their report, the Nuclear Civil Engineering Committee highlighted the 11 faults on the map designated as level I and concluded that only one could be verified through geological surveys [5]. This impeded the implementation of the tectonic geomorphological methods in the field of seismic design for nuclear power plants. Of those 11 faults, 10 were later deemed certain active faults that the Headquarters for Earthquake Research Promotion declared causative faults. This reveals the scope of errors of judgment in the field of seismic design in relation to nuclear power plants. As this illustrates, issues with research history underlie the division between active fault research and seismic design in nuclear power plants.

A further issue is that stakeholders had not faced up to the limitations of science with humility. Traditionally, there was an implicit presumption that active faults would become apparent as long as detailed surveys were conducted. It was also assumed that if active faults were discovered, the scale of the displacement and the magnitude of the earthquakes they might cause could be understood definitively. The established story was that as long as gaining understanding in this manner was possible, stakeholders could discuss whether buildings would be safe in the event that an active fault activated. It should be noted that some active fault researchers supported this line. Under these assumptions, however, it would mean that active faults were considered absent if surveys did not reveal concrete evidence indicating their presence. In addition, an illusion of safety could be created from the misconception that appropriate mitigation measures could be adopted if an active fault was present, because its activity could be predicted. Such disregard for the scientific limitations ultimately invited underestimation of active faults.

Active fault research aims to predict the displacement of active faults. Personally, I conduct my research in the belief that the contribution to the field of engineering and society is also important [6]. Researchers tend to publicize what can be predicted, but accepting the limitations of prediction and making sincere efforts to face facts are essential. Gray areas in research should never be declared "white," i.e., that all is clear.

2.2 Issues with Research and Review Frameworks

Issues with research and review frameworks constitute another reason why active faults have been underestimated in Japan.

The first issue is that active fault surveys for nuclear power plant construction are conducted not during the site review stage when the construction site is determined, but later at the seismic safety review stage when details on facility layout and durability are determined. Predicting the degree of shaking and estimating how a fault might diverge to rupture the surface with sufficient accuracy is difficult for sites very close to active faults. Thus, these sites should be avoided; however, since active fault surveys are conducted after the construction site has been determined, the results of active fault surveys cannot contribute to construction site selection. Clearly, active faults should be investigated as part of the process to determine a suitable construction site. However, with the current situation, the implicit desire not to find active faults can come into play.

The second issue is that surveys are conducted solely by nuclear power plant operators and they are not verified by independent third parties. Electric power companies employ few experts on active faults and earthquakes (the Central Research Institute of the Electric Power Industry, co-funded by the electric power companies, might take on this role at times). Judgments about whether a fault is active generally depend on the survey plan, which requires a high level of expertise to develop. Mistakenly excavating for a geological survey just a few meters from the most appropriate site could lead to an active fault being overlooked. Without high levels of expertise and a proactive attitude, evidence of active faults could simply remain undiscovered. In addition, survey results are rarely made public, although they should be verified. There has also been the suggestion that plant operators have disregarded inconvenient external opinions.

The final issue is that of the review framework. Reviews by the former Safety Agency and the former Nuclear Safety Committee generally only involved checking reports written by the operators, i.e., review board members rarely verified the raw data. Concerns are occasionally raised about whether the conclusions of a plant operator are logically sound and whether their explanations are sufficiently thorough. However, serious debate about the determination of facts is rare. Operators tend to present data with the conclusions first, with further explanations supporting their main argument. Thus, although review board members should scrutinize data and documents themselves in order to identify mistakes, past review frameworks did not require such diligence. Furthermore, as mentioned previously, the fact that review board members did not necessarily adhere to the Revised Regulatory Guide 2006 and Review Guidelines 2010 was also a problem; however, where the responsibility for this lies remains unclear. Moreover, the selection criteria for review board members are vague and members tend to limit to a certain handful of long-serving experts. The reasoning seems to be that active faults cannot be identified appropriately without deep knowledge of conditions applicable to nuclear power plants. Clearly, knowledge of nuclear power plants is unnecessary in objectively assessing the existence of active faults. If anything, this would disrupt neutrality. Many active fault researchers including myself are also responsible for this, having unintentionally become distanced from the problem of nuclear power plants.

The official report of the Fukushima Nuclear Accident Independent Investigation Commission by the National Diet of Japan highlighted inappropriate ties between operators such as the electric power companies and the review framework. It cannot be denied that such structural systems inhibited stricter assessment of active faults. Even to this day, some experts remain steadfast against making assessments stricter. After 2006, however, the numbers of review board members in both the former Safety Agency and the former Nuclear Safety Commission increased, which allowed sharp criticisms to be voiced. Nevertheless, there have been few cases where operators have confirmed the presence of an active fault following criticism from board members. In some nuclear power plants, back-checks have not been concluded because additional surveys intended to negate claims by operators were prolonged endlessly. In some cases, the discussion was terminated before board members felt fully satisfied.

2.3 Problems with Review Rules

There have also been issues with the regulations for nuclear power plant safety reviews in relation to active faults.

The first issue is the problem with the Regulatory Guide for Reviewing Nuclear Reactor Siting Evaluation and Application Criteria, as mentioned previously. This states that, as a general siting rule, "(1) In addition to the site having no history of phenomena that could cause major accidents, it must have no possibility of such phenomena occurring in the future." However, it is unclear whether the presence of an active fault within the site conflicts with the regulatory guide.

The second problem concerns issues with the revised Regulatory Guide for Reviewing Seismic Design, produced in 2006. As a general rule, it outlines that "Facilities that are significant in terms of seismic design … must be designed so that its safety function will not be compromised in the face of seismic force from earthquake motion that is extremely unlikely from a seismic engineering standpoint but still has potential to occur, and in which anticipating significant effects on facilities would be appropriate." Similar statements were also written for tsunamis.

A pertinent question is whether the Fukushima Daiichi nuclear disaster violated the regulations according to these statements. Of specific interest is the wording, "from a seismic engineering standpoint … anticipating significant effects … would be appropriate." This could alternatively be interpreted to mean that, unless deemed that "anticipating significant effects would be appropriate," there would be no obligation to prepare any countermeasures. The body or organization responsible for determining what is appropriate is also unspecified. This clause should have made strict countermeasures mandatory by stating straightforwardly, "… must be designed so that its safety function will not be compromised in the face of seismic force that is extremely unlikely but still has potential to occur."

The third issue is the problem with the "Guidelines for Reviewing Seismic Safety Concerning Geological Faults" published in June 2008 and renewed in 2010 as the "Guidelines for Reviewing Seismic Safety of Nuclear Power Reactor Facilities" (Review Guidelines 2010). In this publication, lessons learned from the oversight of active faults were considered of paramount importance and the guidelines contained strict instructions. The following points for consideration when identifying "active faults that should be examined for seismic design" were specified: (1) Judgments should take into consideration the accuracy and validity of survey results. Moreover, judgments should err on the conservative, if any survey method indicated the possibility of the presence of an active fault that should be examined for seismic design. (2) If cumulative crustal movement since the late Pleistocene cannot be denied, and if fault movement cannot be ruled out as the cause, based on the use of appropriate crustal movement models, the presence of active faults must be assumed appropriate.

If stakeholders had adhered to the above, oversight in the assessment of active faults should practically have been impossible. However, it appears that "conservative judgments" were rarely made. The underlying reasons are that there were no criteria for what constitutes a "conservative judgment" and that the meaning of the word "appropriate" is subjective. The word "appropriate" in this context is rather unnecessary; instead, the clause should simply have stated, "the presence of active faults must be assumed." In addition, unless the clause clarifies that the judgment of "appropriateness" is at the reviewers' discretion, this could invite barren controversy between plant operators and reviewers.

As mentioned previously, with regard to how important establishments should not be sited over active faults, the statement only expressed that such scenarios would be "unanticipated." This was also problematic because the assignment of responsibility remained vague, if such as scenario were to be presented.

References

1. Suzuki Y (2013) Nuclear power plants and active faults – "unanticipated" is unacceptable. Iwanami Shoten, Tokyo (in Japanese)
2. Matsuda T (1975) Magnitude and recurrence interval of earthquakes from a fault. Zishin 2(28):269–283. (in Japanese)
3. The Research Group for Active Faults of Japan (1980) Active faults in Japan: sheet maps and inventories. University of Tokyo Press, Tokyo. (in Japanese)
4. The Research Group for the Active Faults of Japan (1980) Active faults in and around Japan: the distribution and degree of activity. J Nat Dis Sci 2:66–99
5. Japan Society of Civil Engineers, Nuclear Civil Engineering Committee (1985) Ground assessment methods for nuclear power plant. Doboku Gakkai Ronbunshu 358:11–26. (in Japanese)
6. Suzuki Y, Nakata T, Shimazaki K (1999) Geomorphological/geological challenge to the long-term seismic hazard evaluation of active faults. Soil Mech Found Eng 47(2):5–8. (in Japanese)

Chapter 3
Actual Cases in Which Active Faults Were Underrated

Abstract Indications regarding the underestimation of active faults began in 2002 at the Shimane Nuclear Power Plant. Initially, these indications were neglected, but the situation changed significantly in 2006 when active fault researchers proved the presence of a new active fault. The revision of the Regulatory Guide in 2006 highlighted previous assessment issues and reinforced the assessment of active faults. The Noto Peninsula and Niigata Chuetsu-oki earthquakes that occurred in 2007 revealed an underestimation of active faults in the seismic assessment of the Shika and Kashiwazaki-Kariwa nuclear power plants, respectively. Similar problems were subsequently raised for Tsuruga, Higashidori, and Oma power plants.

Keywords Underestimation of active faults · Nuclear power plants · Seismic assessments · Tectonic geomorphology · Safety review guidelines

3.1 The Kashima (Shinji) Fault near the Shimane Nuclear Power Plant, Shimane Prefecture

The first case in which underestimation of active faults near a nuclear power plant was found concerned the Shimane Nuclear Power Plant operated by the Chugoku Electric Power Company. In 2002, Dr. Takashi Nakata surveyed this region and discovered an 18-km-long active fault [1]. However, the operator claimed that the length of the fault was only 8 km (later changed to 10 km). Both the former Safety Agency and the former Nuclear Safety Commission accepted this explanation and they granted approval for the construction of Unit 3 in 2005 (Fig. 3.1).

Fault length and earthquake magnitude are generally thought to be correlated, with a 10-km-long fault estimated capable of causing an M6.5 earthquake (estimates differ in the present day, because the magnitude of earthquakes caused by shorter faults was later revised upward). As the maximum magnitude of such an

Y. Suzuki, *Active Faults and Nuclear Regulation*,
https://doi.org/10.1007/978-981-15-0765-6_3

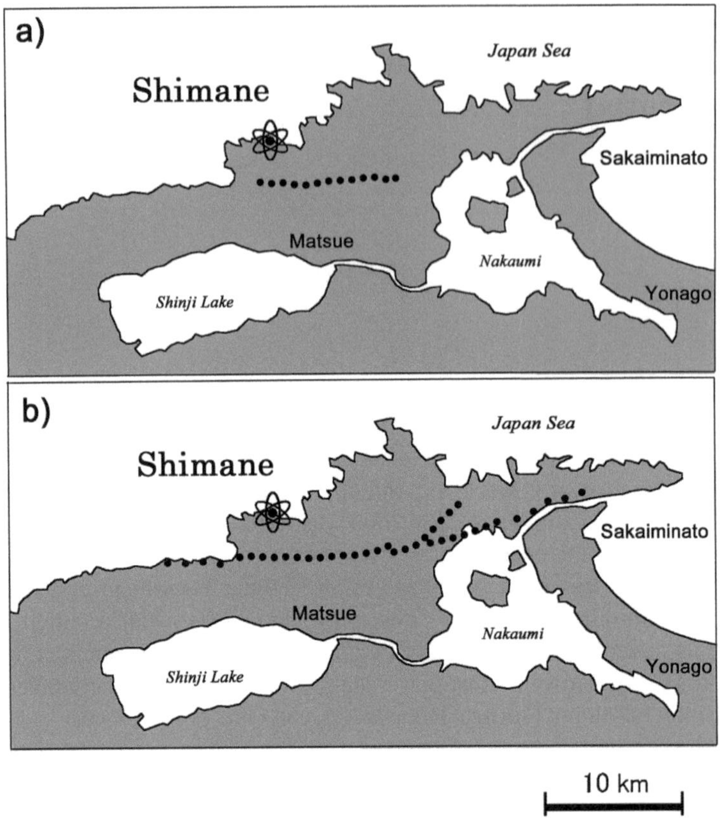

Fig. 3.1 Difference in evaluation of Kashima (Shinji) Fault between that by Chugoku Electric Company (**a**) and by Dr. Takashi Nakata (**b**). Translated from Suzuki [2]@Yasuhiro Suzuki 2013

earthquake was estimated to be M6.5, an M6.5 earthquake should have been anticipated regardless of the presence of an active fault. This means that the presence of an active fault less than 10 km in length would have been irrelevant for seismic design. However, the active fault would have constituted a serious problem if the fault length were actually 18 km. Former Prime Minister Junichiro Koizumi claimed in the House of Representatives on July 29, 2005 (Naikaku-shushitsu 162 No. 100) that the government's interpretation was correct and the research results of Dr. Nakata were criticized as mere speculation based only on geomorphological surveys.

The tables were turned in 2006. Dr. Nakata and colleagues conducted a trench survey at a site where the operator and the government had denied the presence of active faults. The results of the survey indicated very obviously the presence of an active fault. They discovered the Kashima Fault had possibly moved three times over the previous 9000 years, the latest likely being in 880 when the Izumo Earthquake occurred [3]. This, according to the former Regulatory Guide 1978 used

Fig. 3.2 Excavation study for Kashima (Shinji) fault in June 2006. White arrow shows the location of active faults. Reprinted from Suzuki [2] @Yasuhiro Suzuki 2013

for safety reviews before 2005, is what would be called an exceptional active fault requiring special caution. In particular, at that time, facilities relevant to nuclear power plants were ranked in order of importance as class As, A, B, or C. Earthquakes caused by faults with high activity levels, such as those that caused historical earthquakes, were considered S1 Earthquakes (maximum design basis earthquakes). The design of facilities above A class was required to withstand this level of activity. Conversely, active faults with relatively lower levels of activity were considered S2 Earthquakes (extreme design basis earthquakes). If only As class facilities could maintain their safety function, then the design would be considered acceptable. That such exceptional active faults were underrated was an extremely grave issue (Fig. 3.2).

Revisions to the Regulatory Guide were discussed by the Nuclear Safety Standards Subcommittee in 2006. Thus, this fact was treated with gravity and discussions ensued to prevent further neglect of active faults. The subcommittee appears to have cut short the discussion because the revisions had to be rushed; nevertheless, as a result of issues raised by Dr. Katsuhiko Ishibashi (Professor Emeritus, Kobe University) and his colleagues, the full implementation of tectonic geomorphological methods and the need for trench surveys were clearly stated in Revised Regulatory Guide 2006.

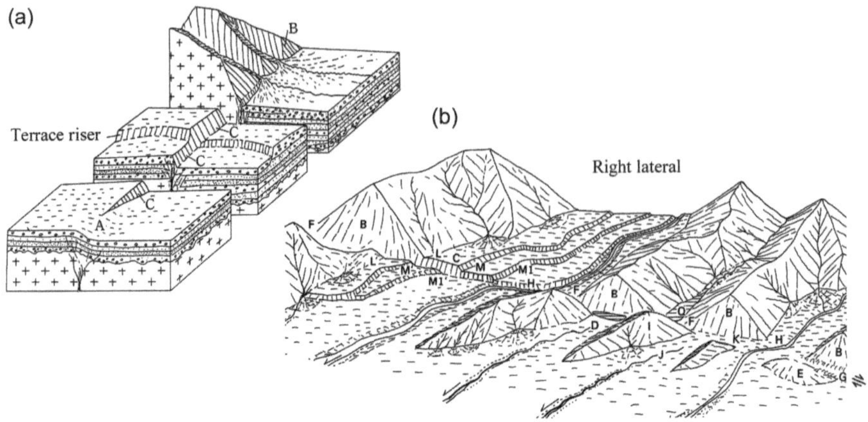

A: Flexure scarp. B:Truncated Spur, C:Fault scarplet, D:Sag pond, E:Tectonic bulge, F:Fault saddle, G:Fault graben, H:Offset drainage channel, I:Shutter ridge, J:Truncated river channel, K:Wind gap, L-L':Offset piedmont line, M-M':Offset terrace riser, Q:Dammed lake

Fig. 3.3 Schematic figure showing tectonic landforms, drawn by Dr. Atsumasa Okada. Translated from Suzuki [2] @Yasuhiro Suzuki 2013

When active faults become displaced, obvious rupture can appear at the surface and leave a mark on surface landforms. Examples include the Neodani Fault in the 1891 Nobi Earthquake and the Nojima Fault in the 1995 Kobe Earthquake, among many others. Visualizing the traces of these events from aerial photos sheds light on the distribution of active faults on a wider scale. This is the basis of the tectonic geomorphological method, i.e., it reveals the presence of active faults by considering landform genesis. The methodology has been established in academia and its efficacy has been verified internationally [4]. The fields of tectonic geomorphology and earthquake geology have supported active fault research over a long time and thus it can be considered "mainstream" (Fig. 3.3). Conversely, the field of nuclear civil engineering conducts surveys that only suggest the possible presence of active faults by focusing on topographic lineaments. The stance of nuclear civil engineering has been that topographic features are merely for reference and that only geological surveys could identify the presence of active faults. Lineament surveys differ fundamentally from tectonic geomorphological surveys.

A tectonic geomorphological survey first elucidates the geomorphic developmental history of a region. Then, a search for the presence of tectonic landforms due to fault displacement is conducted from a genesis perspective. In other words, the survey detects landform abnormalities that could only be explained by fault activity [5]. Therefore, the presence of active faults can sometimes be revealed without geological surveys. Study results typically indicate whether activity can be determined from geomorphology alone by categorizing faults into two levels: "certain" and "inferred," based on the level of certainty of fault activity. However, delay in the transfer of knowledge and technology from the field of tectonic geomorphology to the surveying industry resulted in the method being substituted by lineament surveys. This is why Dr. Nakata was able to discover the Kashima Fault, whereas the plant operator was not [6]. This is a fact that cannot be ignored.

After the Regulatory Guide was revised in 2006, an inspection called a back-check, in which the safety of existing nuclear power plants was examined according to the Revised Regulatory Guide 2006, was conducted. The back-check at Shimane Nuclear Power Plant led to a revision of the fault length to 22 km. However, the only explanation provided was that the length was changed based on the revisions in the Revised Regulatory Guide 2006, and investigation into the reasons why the length was originally underestimated was not conducted [7].

3.2 Active Sea-Bottom Faults near the Shika Nuclear Power Plant, Ishikawa Prefecture

An active sea-bottom fault in the ocean area near the Shika Nuclear Power Plant (operated by the Hokuriku Electric Power Company) was underrated. The Noto-Hanto Earthquake, which occurred on the fault in March 2007, caused vigorous shaking at the site that exceeded predictions (Fig. 3.4). The active sea-bottom fault in question was around 20 km in length and it was located about 18 km northwest of the Shika Nuclear Power Plant. Its presence had been verified by acoustic profiling surveys conducted by the operator when building the Shika Nuclear Power Plant. However, the operator reported it as three short fault segments, each around 6–7 km in length, and it was claimed that the three faults would not move concurrently at the time of an earthquake. As mentioned previously, the need for caution was not necessarily a requirement for a fault shorter than 10 km.

The operator's claim that the fault consisted of three segments was also published as an academic paper [8]. The evidence cited in the paper was that past activity patterns determined by sonic profiling differed among the three segments. However, this was only one possible interpretation and the evidence could not deny the possibility that the faults might move synchronously in the future. The earthquake in 2007 revealed the original conclusion inappropriate and it taught us the lesson that we should not rely on a single possible interpretation.

The Noto Peninsula is known to have many active faults distributed around the area. The Togi-gawa Nangan Fault, stretching approximately in the east–west direction, lies 9 km north of the Shika Nuclear Power Plant. The height distribution of terrace surfaces differs considerably on either side of this fault. From this evidence, Dr. Yoko Ota (Professor Emeritus, Yokohama National University) reported that this fault had shown activity in the late Pleistocene [9]. Furthermore, Dr. Mitsuhisa Watanabe (Professor, Toyo University) highlighted that this fault might extend to the ocean bottom and that it could be causing uplift at the western coast of the Noto Peninsula, where the Shika Nuclear Power Plant stands [10]. On the coast, there are several flat layered topographic surfaces with elevations of several

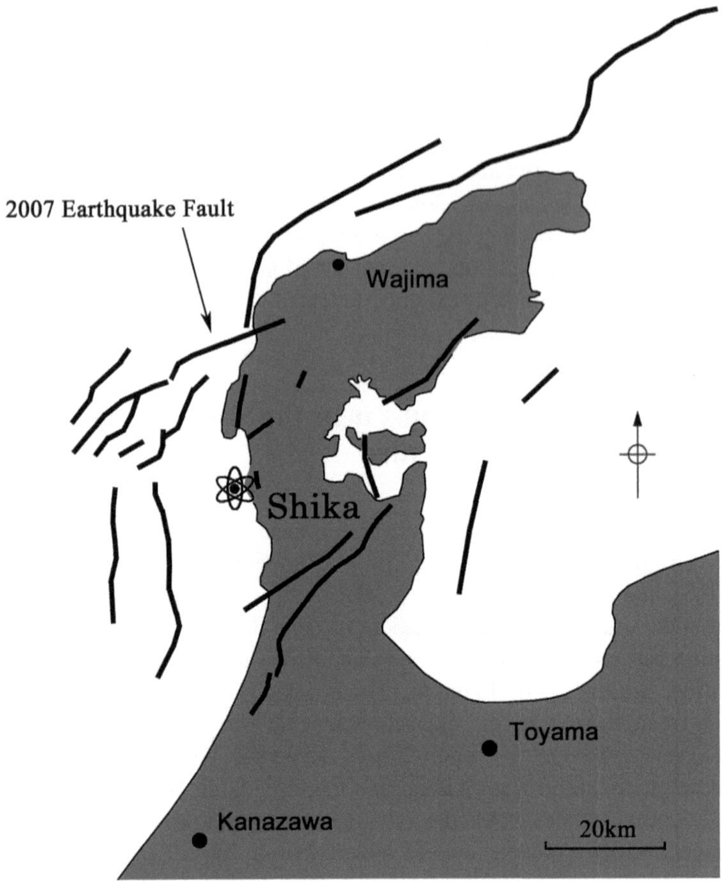

Fig. 3.4 Active faults around Noto Peninsula. (Data from Hokuriku Electric Company). Translated from Suzuki [2] @Yasuhiro Suzuki 2013

meters, which are thought to have been formed by seismic uplift over the previous several thousand years. In Japan, the sea level generally rose several meters during the warm period about 6000 years ago. As levels have been reasonably stable since, it is impossible that multiple terraced landforms could have developed in the coastal area after this period if the crust were stable. The presence of such landforms is evidence that intermittent uplift has occurred because of earthquakes. As of June 2013, the plant operator had not accepted the existence of the Togi-gawa Nangan Fault, or the possibility of coastal uplift due to active faults on the coast. Furthermore, in 2012, the former Safety Agency reported that there was an active fault inside the nuclear plant site, which might have activated within the previous several tens of thousands of years. It is the responsibility of the NRA to verify this claim.

3.3 Active Sea-Bottom Fault Offshore of the Kashiwazaki-Kariwa Nuclear Power Plant, Niigata Prefecture

The Niigata-ken Chuetsu-oki Earthquake occurred in June 2007. TEPCO's Kashiwazaki-Kariwa Nuclear Power Plant was near the epicenter and it suffered extensive damage, such as significant ground deformation by liquefaction within the site. However, for six months after the occurrence of the earthquake, basic information such as the location of the earthquake source fault and its angle remained undetermined. Various hypotheses were proposed based on seismic observational and crustal movement data. We reviewed sonic profiling records of the surrounding oceanic region, which were made publicly available as part of the nuclear power plant license application. We realized that active sea-bottom faults had been overlooked when approval was granted (Fig. 3.5). This was reported in newspapers on July 20, 2007. Subsequently, we obtained analysis records from the operator's sonic profiling survey and we discussed why this had been overlooked in an attempt to prevent future similar occurrences [11].

In essence, the reason why the active sea-bottom faults were overlooked was that the operator focused on whether faults were visible in the sonic profiling; they concluded faults were absent if they did not show. Sonic profiling was designed to track how geological strata connect but not to identify disconnectivity with any certainty. Relatively minor disconnectivity might disappear through repeated rounds of data analysis. Therefore, it is inappropriate to disregard such limitations and to conclude that faults are absent when clear disconnectivity does not appear. The important consideration is not simply whether a fault is visible in the profile, but whether its activity might have caused downthrows or upward flexures. In addition, intense (low-wavelength) deformation indicates that a fault extends directly underneath, even if it does not reach the seafloor.

When examining the fault length immediately after the earthquake with the above in mind, the length was found to extend for at least 35 km. This fault is located on the eastern edge of the Sado Ocean Basin and it is a reverse fault with the east side upthrown. The fault plane slopes to the east, and the Kashiwazaki-Kariwa Nuclear Power Plant is located upward from the fault plane (hanging wall side).

In the back-check discussion that followed, the presence of the active fault in question was reaffirmed. However, there is a feeling that its length had not been discussed adequately. The former Safety Agency accepted the conclusion that the length was 36 km [12], and this corresponds to the length of the fault that caused the current earthquake. However, from a tectonic geomorphology standpoint, it is possible that the length of the active fault including portions that did not move this time could span 50 km, because the Sado Ocean Basin expands further northward [13]. According to this idea, this oceanic region should anticipate an earthquake more intense than the current one, i.e., maximum M7.5 [14]. This is an important point that should be reexamined in future safety reviews. It is crucial to ascertain if the faults were reassessed with strict adherence to the new safety regulations mentioned

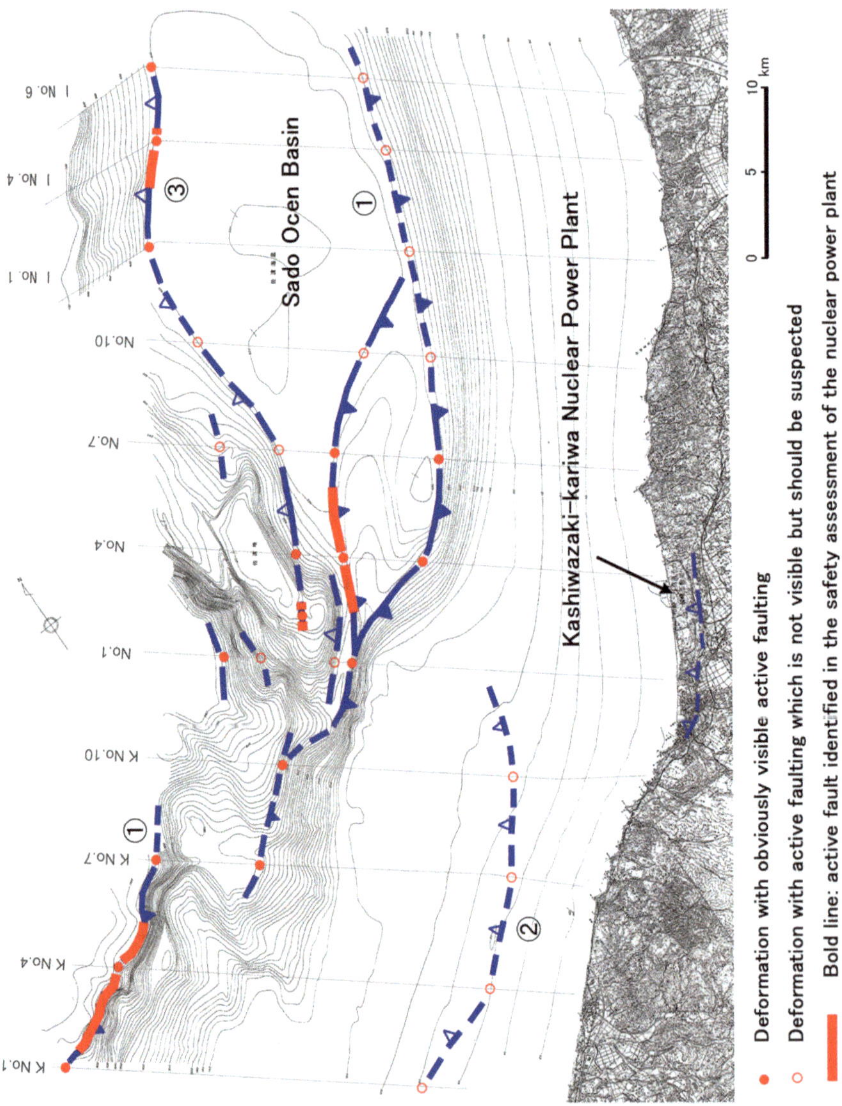

Fig. 3.5 Sea-bottom active faults off Chuetsu area. Black and gray bold lines show active faults. Translated from Suzuki [2] @Yasuhiro Suzuki 2013

in Chap. 5. This leads to debate over whether it is appropriate to declare that the Chuetsu-oki Earthquake (M6.8) is the largest class of earthquake that could possibly occur around the Sado Ocean Basin.

3.4 The Urazoko Fault near the Tsuruga Nuclear Power Plant, Fukui Prefecture

The Urazoko Fault passes through the site of the Tsuruga Nuclear Power Plant operated by the Japan Atomic Power Company. This was classified as a certainty level I active fault in "Active faults in Japan [Revised Edition]" [15] (University of Tokyo Press), published in 1991. Although this was regarded as a certain active fault within academia, the operator did not accept this until 2008. The detailed sequence of events is illustrated in Chap. 4. A trench survey conducted by the operator in 2008 revealed clear evidence of activity (Fig. 3.6).

Fig. 3.6 Excavation for Urazoko Fault in 2008. Reprinted from Suzuki [2] @Yasuhiro Suzuki 2013

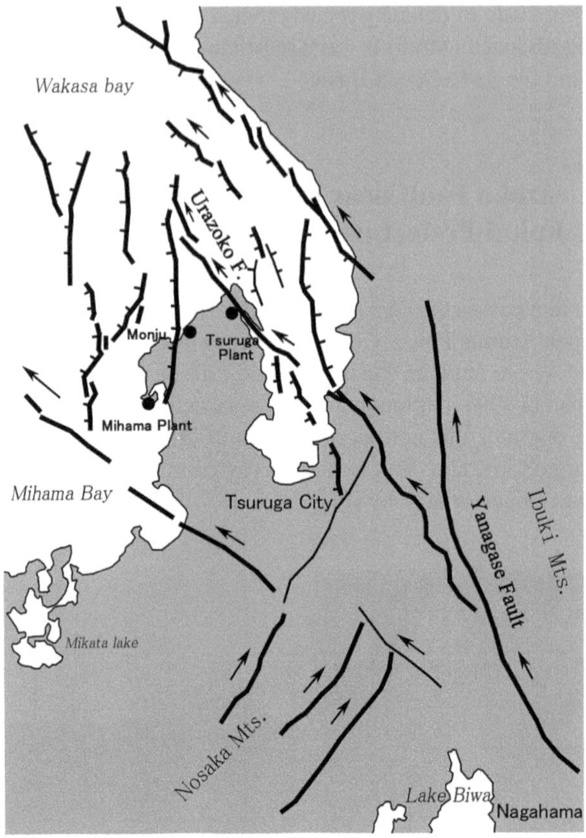

Fig. 3.7 Active faults around Tsuruga city. Arrow shows the slip direction, simplified from Sugiyama et al. [16]. Translated from Suzuki [2] @Yasuhiro Suzuki 2013

The trench was of massive scale, exceeding 15 m in depth. A fault plane that clearly ruptures new strata near the surface was also found on its wall surface. According to Dr. Yuichi Sugiyama (National Institute of Advanced Industrial Science and Technology) and colleagues, this fault had shown activity twice in the previous 7000 years. The most recent event occurred in the previous 4500 years; when accounting for lateral slips, the slip amount during a single activity event was over 3 m and the fault length spanned 35 km [16]. Compared with the values determined by the operator in 2012, these values were 1.4 times longer in length and nearly 2 times larger in slip amount (Fig. 3.7).

The Urazoko Fault is a reverse fault with uplift on its northeastern side and clear left-lateral offset. It moves in accordance with the stress field of the present day, with compression in the east–west direction. Aerial photos show a mountain range extending in the northwest–southeast direction to the east of the Tsuruga Nuclear Power Plant, with the Urazoko Fault passing through its base. On a wider scale, the mountain range was formed from fault activity, while the western side subsides to

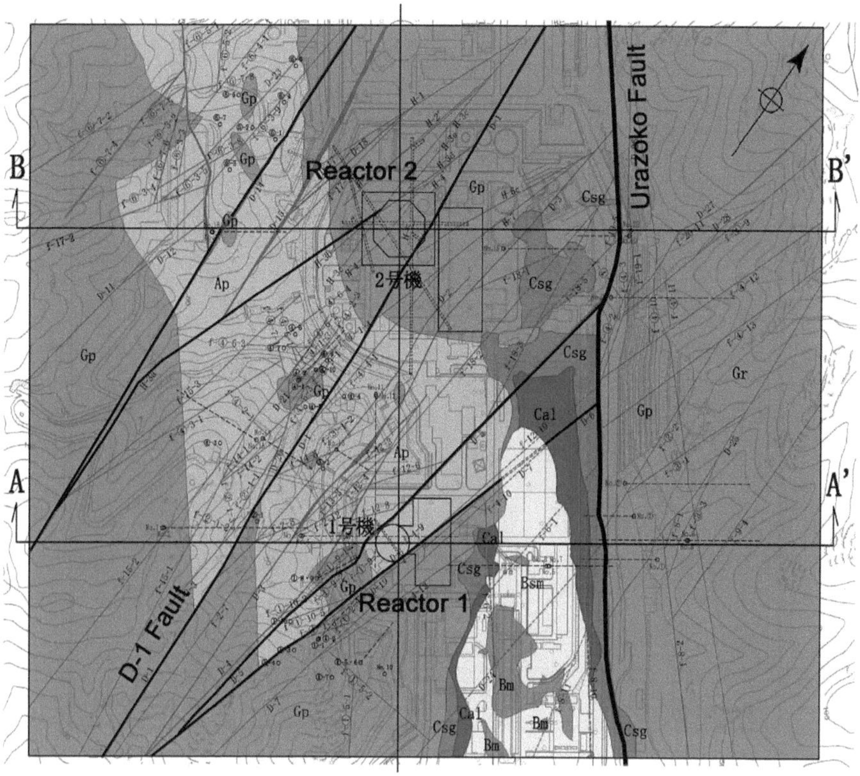

Fig. 3.8 Geological Map of Tsuruga Nuclear Power Plant premise, with data from Japan Atomic Power Company

form the small plains on which the Tsuruga Nuclear Power Plant stands. The trace of the fault, however, is not just along the mountain front but also several tens of meters eastward (i.e., the mountain side), where a clear left-lateral offset landform of small streams is present. In tectonic geomorphological terms, this trace would be considered important. However, misinterpretation of the lineament survey suggested the possible fault was located only along the mountain front; thus, attention was not focused on the right trace, which might have led to the active fault being overlooked.

The nuclear reactor at the Tsuruga Nuclear Power Plant is only 250 m from the Urazoko Fault. Experts have expressed concern about this, asking, “Should nuclear power plants be standing so close by?” and stating that “If they knew of the existence of the active fault, they would not have built the plant.” They say this because it is generally difficult to predict accurately what might happen near an active fault when an earthquake occurs. Geological maps of the area show numerous faults close by and linked to the Urazoko Fault. Of crucial importance is whether some of these faults might move in association with movement of the Urazoko Fault. This contributed to the problem of awareness in the fault surveys conducted within the Tsuruga Nuclear Power Plant site after 2012 (Fig. 3.8).

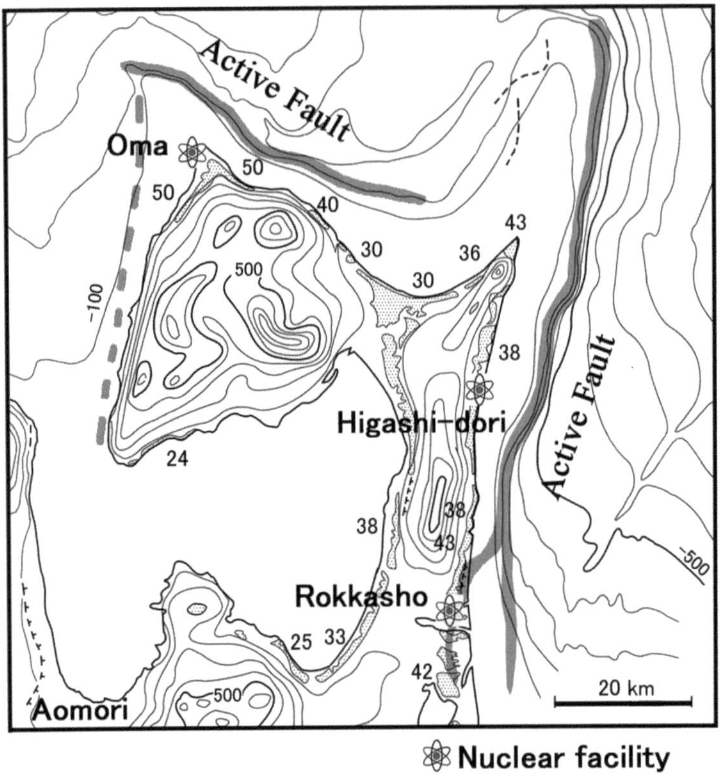

Fig. 3.9 Active faults around the Shimokita Peninsula. Figure is presented by courtesy of Dr. Mitsuhisa Watanabe

3.5 Active Sea-Bottom Faults near the Shimokita Peninsula, Aomori Prefecture

The Shimokita Peninsula houses the Higashidori Nuclear Power Plant operated by the Tohoku Electric Power Company, TEPCO's Higashidori Nuclear Power Plant (under construction), and the Rokkasho Village Nuclear Fuel Cycle Facility operated by Japan Nuclear Fuel Limited. In addition, Japan Nuclear Fuel Limited plans to build a new nuclear power plant in Oma. Many active sea-bottom faults exist beneath the coastal waters of the Shimokita Peninsula. Although some are thought to continue onto land areas, to date, they have rarely been taken into account in seismic designs.

In the area off the Shimokita Peninsula, there is a clear fault spanning 100 km along the continental-shelf margin. Although some asserted in the 1990s that this should be classified as active, this was refuted. Dr. Yasutaka Ikeda (University of Tokyo) independently analyzed reflection survey results and asserted that this was indeed an active fault [17]. As this is a reverse fault, in which the fault plane dips

Fig. 3.10 Benten Island off Oma with co-seismic uplifted coastal benches. Photo is reprinted by courtesy of Dr. Mitsuhisa Watanabe

toward the west, both the Higashidori Nuclear Power Plant and the Rokkasho Village Nuclear Fuel Cycle Facility are located on the hanging wall side (the upper side of the fault plane). Movement of a fault along the continental-shelf margin would cause intense shaking. Furthermore, the hanging wall side would also become a deformation zone, making it more likely for associated faults in the coastal area to move synchronously (Fig. 3.9).

Dr. Mitsuhisa Watanabe found that the Rokkasho Fault, which diverges from the fault along the continental-shelf margin, lies directly underneath the Rokkasho Village Nuclear Fuel Cycle Facility, and that it is causing flexure on the geomorphic surface close to the facility [18]. Deformation of strata can also be inferred from reflection survey records. Meanwhile, numerous faults are distributed within the site of the Higashidori Nuclear Power Plant, and the NRA was recently investigating whether any of these faults are active. The assessment meeting in May 2013 concluded that many of the faults within the site are active faults that should be examined for seismic design.

Similar to the Shika Nuclear Power Plant, evidence of seismic uplift was confirmed near the site of Oma Nuclear Power Plant. A bench formed very recently (a flat topographic surface formed by coastal erosion) has been uplifted with elevation of several meters. This formation is prominent on Benten Island, 1 km off Cape Oma. The height distribution of such geomorphological features increases toward the north. This implies that they have been uplifted intermittently with tilting under the influence of the activities of active sea-bottom faults located northern offshore region. Furthermore, multiple coastal terrace surfaces formed in the previous several tens of thousands of years are prominent on land. Their height distribution should originally have been planar; however, it is currently higher toward the north, which also strongly implies active sea-bottom fault movement [19] (Figs. 3.10 and 3.11).

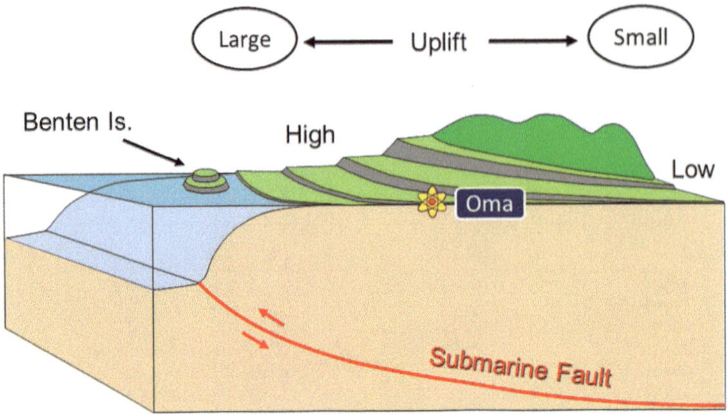

Fig. 3.11 Schematic figure of tectonic landforms around Oma. Figure is presented by courtesy of Dr. Mitsuhisa Watanabe

References

1. Sato T, Nakata T (2002) Kashima fault as a model fault for active fault segmentation. Active Fault Res 21:99–110. (in Japanese)
2. Suzuki Y (2013) Nuclear power plants and active faults – "unanticipated" is unacceptable. Iwanami Shoten, Tokyo (in Japanese)
3. Watanabe M, Nakata T, Okumura K, Kumahara Y, Goto H, Kumamoto T, Imaizumi T, Tokuoka T, Fukita A (2006) Excavation study of Kashima Fault, Programme and Abstracts. Seismological Society of Japan Fall Meeting, 50. (in Japanese)
4. Suzuki Y (2013) Tectonic geomorphological active fault studies in Japan after 1980. Geogr Rev Jpn, Series B 86:6–21
5. Watanabe M, Suzuki Y (1999) Aerial photograph interpretation for active fault identification, Kokon Shoin. (in Japanese)
6. Nakata T (2009) Why important tectonic geomorphological method is for active fault research. Kagaku 79:167–174. (in Japanese)
7. Nakata T (2011) Inaccurate assessment on active fault in nuclear regulation. Asahi J 116(25):33–35. (in Japanese)
8. Katagawa H, Hamada M, Yoshida S, Kadosawa H, Mitsuhashi A, Kono Y, Kinugawa Y (2005) Geological development of the West Sea area of the Noto Peninsula District in the Neogene tertiary to quaternary, Central Japan. J Geogr 114:791–810. (in Japanese)
9. Ota Y, Matsuda T, Hirakawa K (1976) Active faults in Noto Peninsula, Central Japan. Quatern Res 15:109–128. (in Japanese)
10. Watanabe M, Suzuki Y (2012) Co-seismic uplift and active fault around in the western part of Noto Peninsula, Central Japan, Japan Geoscience Union Meeting, HSC24–11
11. Suzuki Y, Nakata T, Watanabe M (2008) Essential problem of active fault assessment in safety regulation of nuclear power plants in Japan. Kagaku 78:97–102. (in Japanese)
12. Okamura Y (2010) Active structure of the 2007 Chuetsu-oki earthquake source area and the Sado Basin, Central Japan. Active Fault Res 33:15–25. (in Japanese)
13. Watanabe M, Nakata T, Suzuki Y (2010) Active submarine faults along the eastern margin of the Sado Basin near the focal area of the 2007 Chuetsu-oki earthquake, Northeast Japan. Active Fault Res 33:27–37. (in Japanese)
14. Ishibashi K (2009) Allowance of extremely underestimated earthquakes in Kashiwazaki-Kariwa nuclear plant assessment. Kagaku 79:9–17. (in Japanese)

15. The Research Group for Active Faults of Japan (1991) Active faults in Japan: sheet maps and inventories, Revised edn. University of Tokyo Press, Tokyo. (in Japanese)
16. Sugiyama Y, Sakamoto I, Takino Y, Usami T, Kishimoto H, Nagata T (2012) Examination on geometry, size and rupturing history of the Urazoko-Yanagaseyama fault zone. Ann Rep Active Fault Paleoearthquake Res 12:149–193. (in Japanese)
17. Ikeda Y (2012) An active fault at the continental-shelf margin off the Shimokita Peninsula: how did the nuclear safety commission evaluate seismic hazards from the large-scale fault extending down-dip beneath nuclear facilities? Kagaku 82:644–650. (in Japanese)
18. Watanabe M, Nakata T, Suzuki Y (2009) Rokkasho fault underlying the nuclear fuel cycle facility. Kagaku 79:182–185. (in Japanese)
19. Watanabe M, Nakata T, Suzuki Y, Koiwa N (2012) Co-seismic uplift and submarine active fault around the northwestern end of Shimokita Peninsula, Northeast Japan. Active Fault Res 36:1–10. (in Japanese)

Chapter 4
Active Fault Surveys at Nuclear Power Plant Sites: Actions Taken by the NRA and Their Impact

Abstract In April 2011, the former Nuclear and Industrial Safety Agency indicated that the active faults at the sites of nuclear power plants might not have been thoroughly investigated. If an active fault is present directly under an important facility such as a nuclear reactor, it would violate the rules of the nuclear safety review at that time. For this reason, operators were requested to investigate this aspect immediately. The Nuclear Regulation Authority (NRA), launched in September 2012, took over this survey and began the compilation of a new safety review. The author participated in this investigation and examination as an outside expert and recognized the existence of an active fault at the Tsuruga Nuclear Power Plant site. In the examination, intense discussions were carried out repeatedly with the operator. In May 2013, NRA confirmed the existence of active faults at the Tsuruga Nuclear Power Plant site. All discussions were made public and distributed on the Internet. The media reported this issue every day and evaluated the conclusions of the NRA.

Keywords Active faults · Nuclear power plant sites · Nuclear and Industrial Safety Agency · Nuclear Regulation Authority · Evaluation by media

4.1 Problem Presentation by the Former Nuclear and Industrial Safety Agency

The earthquake that occurred on April 11, 2011, in Hamadori, Fukushima Prefecture, revealed a considerable problem with active fault assessments during nuclear power plant reviews. This earthquake occurred because of recurring movement in active faults, which caused obvious seismic faults to appear on the surface [1]. However, some of these had been deemed inactive in the assessment undertaken during the license application process for TEPCO's Fukushima Daiichi Nuclear Power Plant. The fact that this fault was a normal fault with low probability of activity given

Y. Suzuki, *Active Faults and Nuclear Regulation*,
https://doi.org/10.1007/978-981-15-0765-6_4

today's stress field was given considerable weight in the construction review of the former Nuclear Safety Commission in June 2010. However, since this normal fault did move, it became clear that considering normal faults as not active is inappropriate. This lesson was observed with gravity, and on April 29, 2011, the former Safety Agency required operators across the nation to reassess faults and the geology around nuclear power plants and to report the outcomes as soon as possible.

After September 2011, to hear professional opinions regarding earthquakes and tsunamis, the former Nuclear and Industrial Safety Agency (MISA) hosted a meeting at which they began discussion both on coupled movement in active faults and on assessment problems of tsunamis. In addition, after April 2012, they opened discussion on faults within nuclear power plant sites. They visited the Tsuruga Nuclear Power Plant on April 24 and they ordered the Japan Atomic Power Company to conduct a full reexamination, because there was a possibility that a fault near the nuclear reactor building might have moved in the new geologic era. Examination targets included faults that passed underneath the nuclear reactor. They explained that the nuclear power plant site would be deemed inappropriate unless the possibility of this fault moving in association with active faults in the area could be eliminated. This was the first of a series of onsite fault surveys (Fig. 4.1).

Note, however, that although surveys of faults underneath nuclear reactors appear sensational, this represents only the tip of the iceberg in relation to the active fault problem at nuclear power plant sites. Additionally, some even go so far as to say that this is of low priority, because the faults are often of relatively small scale.

4.2 Organizing the Nuclear Regulation Authority Survey Team

Immediately upon being established in September 2012, the NRA began conducting onsite fault surveys, carrying on the work of the former Safety Agency. As mentioned previously, this issue was only the tip of the iceberg in relation to the problem of active faults near nuclear power plants. Nevertheless, this was handled seriously as a predetermined policy.

In late September 2012, in undertaking the onsite surveys, the NRA called for the support of four academic societies that undertake active fault research: the Japanese Society for Active Fault Studies, Japan Association for Quaternary Research, Geological Society of Japan, and Seismological Society of Japan. The NRA requested the societies nominate survey member candidates. When doing so, they added the condition that candidates must not have been involved in previous active fault assessments of individual plants, such as with the former Nuclear Safety Commission and the former Safety Agency. This stipulation was made not because of issues with former review board members but, instead, to refresh the atmosphere such that an objective review free of preconceptions could be facilitated.

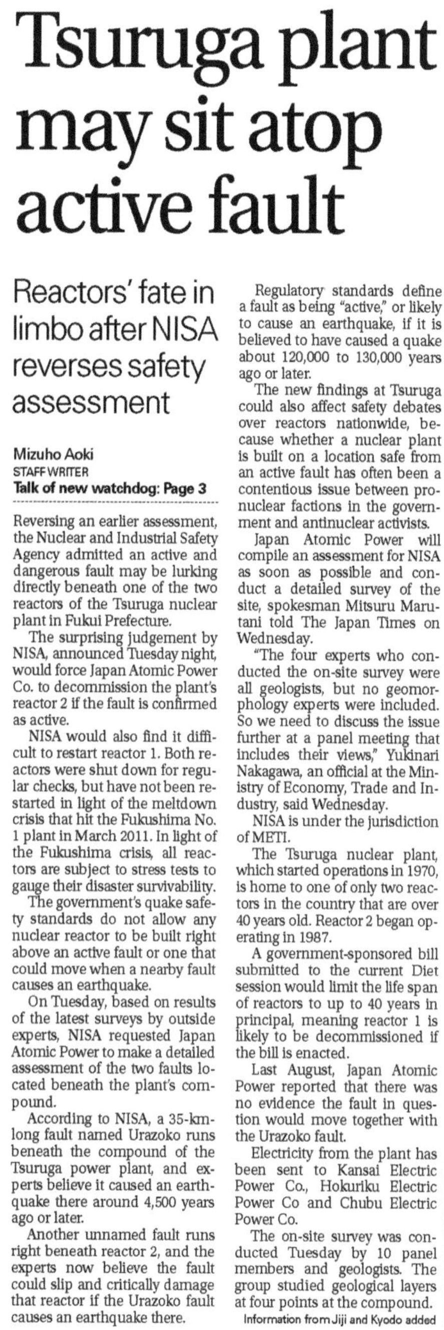

Tsuruga plant may sit atop active fault

Reactors' fate in limbo after NISA reverses safety assessment

Mizuho Aoki
STAFF WRITER
Talk of new watchdog: Page 3

Reversing an earlier assessment, the Nuclear and Industrial Safety Agency admitted an active and dangerous fault may be lurking directly beneath one of the two reactors of the Tsuruga nuclear plant in Fukui Prefecture.

The surprising judgement by NISA, announced Tuesday night, would force Japan Atomic Power Co. to decommission the plant's reactor 2 if the fault is confirmed as active.

NISA would also find it difficult to restart reactor 1. Both reactors were shut down for regular checks, but have not been restarted in light of the meltdown crisis that hit the Fukushima No. 1 plant in March 2011. In light of the Fukushima crisis, all reactors are subject to stress tests to gauge their disaster survivability.

The government's quake safety standards do not allow any nuclear reactor to be built right above an active fault or one that could move when a nearby fault causes an earthquake.

On Tuesday, based on results of the latest surveys by outside experts, NISA requested Japan Atomic Power to make a detailed assessment of the two faults located beneath the plant's compound.

According to NISA, a 35-km-long fault named Urazoko runs beneath the compound of the Tsuruga power plant, and experts believe it caused an earthquake there around 4,500 years ago or later.

Another unnamed fault runs right beneath reactor 2, and the experts now believe the fault could slip and critically damage that reactor if the Urazoko fault causes an earthquake there.

Regulatory standards define a fault as being "active," or likely to cause an earthquake, if it is believed to have caused a quake about 120,000 to 130,000 years ago or later.

The new findings at Tsuruga could also affect safety debates over reactors nationwide, because whether a nuclear plant is built on a location safe from an active fault has often been a contentious issue between pro-nuclear factions in the government and antinuclear activists.

Japan Atomic Power will compile an assessment for NISA as soon as possible and conduct a detailed survey of the site, spokesman Mitsuru Marutani told The Japan Times on Wednesday.

"The four experts who conducted the on-site survey were all geologists, but no geomorphology experts were included. So we need to discuss the issue further at a panel meeting that includes their views," Yukinari Nakagawa, an official at the Ministry of Economy, Trade and Industry, said Wednesday.

NISA is under the jurisdiction of METI.

The Tsuruga nuclear plant, which started operations in 1970, is home to one of only two reactors in the country that are over 40 years old. Reactor 2 began operating in 1987.

A government-sponsored bill submitted to the current Diet session would limit the life span of reactors to up to 40 years in principal, meaning reactor 1 is likely to be decommissioned if the bill is enacted.

Last August, Japan Atomic Power reported that there was no evidence the fault in question would move together with the Urazoko fault.

Electricity from the plant has been sent to Kansai Electric Power Co., Hokuriku Electric Power Co and Chubu Electric Power Co.

The on-site survey was conducted Tuesday by 10 panel members and geologists. The group studied geological layers at four points at the compound.

Information from Jiji and Kyodo added

Fig. 4.1 Newspaper article of *The Japan Times* on April 26, 2012, reporting assessment result of the fault in the Tsuruga Nuclear Power Plant premise by the Nuclear and Industrial Safety Agency (MISA). Reprinted by permission from *The Japan Times*

The candidates nominated by the four academic societies were proposed to the NRA by October 6, 2012. Subsequently, the NRA requested the participation of selected candidates on the list, and the names of the 16 candidates who accepted the request and became fault survey members were made public on October 17, 2012.

First, they formed a survey team for KEPCO's Oi Nuclear Power Plant. The preliminary meeting and onsite survey were carried out on October 23 and November 2, respectively [2]. NRA Chairperson Shunichi Tanaka had clearly stated that nuclear reactors would be shut down if the likelihood of a fault being active was high (and not only when active fault presence was certain) [3].

I was part of the Tsuruga Nuclear Power Plant survey team, which was formed after the Oi team and carried out onsite surveys on December 1–2, 2012. Survey team leader Kunihiko Shimazaki gave firm instructions as Deputy Chairperson of the NRA, stating, "The survey team's fundamental mission is to make scientific judgments. Do not let social or economic factors influence your thinking. We will be accused if we make non-scientific judgments." It was also confirmed that survey team members should act as independent researchers, not as representatives of academic societies. In science, academic societies tend to have "the search for truth" as their objective. Thus, they respect individual observation and judgment, and they do not attempt to agree on an official opinion for society. The various opinions from members selected through a transparent process were taken with weight as something close to the truth.

Since the Fukushima Daiichi nuclear disaster, the public has come to understand that disasters that occur once in a millennium should not be taken lightly. Thus, ample caution for rare earthquakes caused by active faults is definitely required. The surveys tested whether the NRA could avoid regulatory capture—something for which nuclear power plant governance had been criticized previously—and instead, enforce strict regulations that prioritize safety.

4.3 Unique Circumstances of the Tsuruga Nuclear Power Plant

The Tsuruga Nuclear Power Plant has a peculiar situation in that it has a massive active fault within the site. Operation of the plant had continued for nearly 40 years without consideration that the fault was active. The Tsuruga Nuclear Power Plant filed a license application for Unit 1 in October 1965. This was approved in April 1966 and commercial operation began in March 1970. As this was before a time when active fault research became established, insights were lacking in the field. According to the license application, however, faults underneath the site were surveyed extensively and were considered no longer active [4].

License applications were filed for Tsuruga Unit 2 in March 1979, and they were approved in September 1980. The book *Active Faults in Japan* (University of Tokyo Press), published in the same year, indicated a certainty level III active fault where the Urazoko Fault lies. Subsequently, *Active Faults in Japan* [Revised Edition], published in 1991, raised this fault to certainty level I and named it the "Urazoko Fault." The timings of when these books were published might suggest that insights on active faults had not been obtained in time for the assessment of Unit 2. However, active fault research had fully taken off by 1979; by then, various insights had been realized and suitable survey methods established. Thus, the situation differed markedly from the time when Unit 1 was constructed.

After the license application for Unit 2 was submitted in March 1979, partial revisions were twice made to supplementary documents in August 1980 and April 1981. In particular, the supplementary document revised in 1980 is worthy of specific attention. This document outlined the results of test-pit surveys of faults underneath the nuclear reactor and of trench surveys on the Urazoko Fault. The trench surveys included an outcrop sketch that implied the fault could be active. It is possible that experts checked this at the review stage and requested that additional surveys be conducted. However, the first round of reviews was completed a month later, in September 1980. The survey results did not lead to the identification of this as an active fault, nor was the information utilized in the design.

Later, in March 2004, license applications were filed for Units 3 and 4. When these were submitted, the operator forcibly attempted to deny that the fault was active, even though the results of borehole surveying strongly implied the possibility. In response, the former Safety Agency raised sharp concerns and a trench survey was then undertaken to resolve the disagreement. As decisive evidence was found in 2008, the operator finally accepted the Urazoko Fault officially. The operator did not accept that the fault was active until there was conclusive evidence, even though data strongly implied the possibility. Such an attitude from the very people who should have been prioritizing safety is a grave problem. Moreover, the fact that such scenarios were repeated even after Revised Regulatory Guide 2006 stipulated that "faults in which the possibility cannot be denied will be classified as active" is even more worrisome.

With these lessons in consideration, in conducting the onsite fault survey at the Tsuruga Nuclear Power Plant, we were expected to make judgments by adhering strictly to the regulations from Revised Regulatory Guide 2006 and Review Guidelines 2010. We were obviously also expected to make judgments that considered the fact that the active Urazoko Fault was actually present within the site.

4.4 Process of Onsite Fault Surveys at the Tsuruga Nuclear Power Plant

The onsite fault (and fracture zone) survey team for the Tsuruga Nuclear Power Plant held a preliminary meeting on November 27, 2012. At the meeting, the team agreed that the definition of active faults would follow Revised Regulatory Guide 2006. In other words, active faults that should be examined for seismic design were defined as "faults in which activity after the late Pleistocene cannot be eliminated." The team also reaffirmed that Review Guidelines 2010 outlined, "With regard to active fault identification, judgments should err on the conservative and should take into consideration the accuracy and validity of survey results … if any of the survey methods … indicate possibility of [activity] … judgments should err on the conservative." The reaffirmation of these points bore important meaning when presenting the conclusions of the survey [5].

Onsite surveys were conducted on December 1–2, 2012. Staff in charge on the operator side explained the borehole, trench, and outcrop surveys in detail. Time was also spent confirming other details and on related onsite discussions. Of particular interest was a relatively new fault that passed underneath Unit 2, named D-1 by the operator. The survey team conducted a general inspection on the first day and then focused on specific points of concern raised by team members on the second day. They focused on a fault found through a large-scale borehole (D-1 trench) at a site near where the Urazoko and D-1 faults merge.

The operator focused on a fault later named G Fault and emphasized that it was active only in an older geological period. As this was located near a site assumed to represent an elongated portion of the D-1 Fault, they presented the interpretation that the D-1 Fault itself was of an equivalent age. Their supporting evidence also relied on microanalysis of a portion of the fault that showed scratches from slips and on the fact that both the G and D-1 faults shared a normal-fault-type scratch. However, at a site only 5–6 m from the G Fault, the survey team discovered another fault (K Fault), which clearly ruptured younger geological layers (Fig. 4.2).

The first assessment meeting was held on December 10, 2012. There, the finding that the K Fault cuts through a stratum approximately 150,000 years old (or possibly younger) and that it is very likely active was discussed. There is ample possibility that the K Fault is an extension of the D-1 Fault. As it was difficult to eliminate this possibility, the conclusion was that the D-1 Fault should also be classified as an active fault that should be examined for seismic design. There were no data available at that time that indicated definitively that the K and D-1 faults were connected. However, consideration was taken of the fact that they were oriented in the south–north direction, had dip of 80° to the west, and were only 100 m away from the site confirmed by onsite borehole surveys. The conclusion also reflected the instructions of Review Guide 2010 that "Judgments should err on the conservative and should take into consideration the accuracy and validity of survey results" (Figs. 4.3 and 4.4).

Afterward, however, the team received heightened criticism that they had made one-sided judgments without full consideration of the operator's explanations. Although the first meeting ended with agreement that the assessment report would be completed over e-mail discussions, a second assessment meeting was held on January 28, 2013, where the team reexamined the draft assessment report. In addition, the NRA received expressions of concern over whether it should be acceptable for a small number of experts to make such judgments. Subsequently, a third assessment meeting was held on March 8, 2013. There, the operator's representatives and geological consultants presented new data and offered revised explanations, after

Fig. 4.2 K Fault carefully observed in the investigation at Tsuruga Nuclear Power Plant premise. Photo is reprinted from Suzuki [6]@Yasuhiro Suzuki 2013

which debate ensued. As the survey team still concluded that changes to the assessment would not be sensible, a peer review was conducted by the evening of that day with other experts nominated by the four academic societies. No particular objections to the team's original conclusions were raised through the peer review process. If anything, there were additional concerns raised regarding the reliability of the data presented by the operator.

A five-hour discussion with the operator was conducted at the fourth meeting held on April 24, 2013. As a rebuttal of the findings of the survey team, the operator presented three pieces of evidence. (1) The latest activity exhibited by the D-1 and G faults was normal-fault-type behavior, contradicting the assertion that the K Fault is a reverse fault. (2) As the K Fault does not cut through strata that include volcanic ash from 120,000 years ago, activity after the late Pleistocene can be ruled out. (3) The K Fault extends in a different direction to the D-1 Fault. However, the survey

Fig. 4.3 Excavation site with D-1, K, G, and Urazoko Faults at Tsuruga Nuclear Power Plant premise. Photo is reprinted from Suzuki [6]@Yasuhiro Suzuki 2013

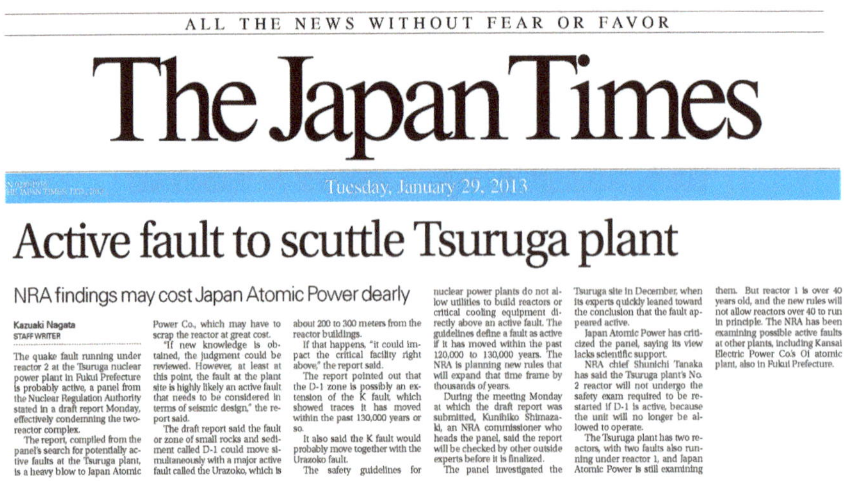

ALL THE NEWS WITHOUT FEAR OR FAVOR

The Japan Times

Tuesday, January 29, 2013

Active fault to scuttle Tsuruga plant

NRA findings may cost Japan Atomic Power dearly

Kazuaki Nagata
STAFF WRITER

The quake fault running under reactor 2 at the Tsuruga nuclear power plant in Fukui Prefecture is probably active, a panel from the Nuclear Regulation Authority stated in a draft report Monday, effectively condemning the two-reactor complex.

The report, compiled from the panel's search for potentially active faults at the Tsuruga plant, is a heavy blow to Japan Atomic Power Co., which may have to scrap the reactor at great cost.

"If new knowledge is obtained, the judgment could be reviewed. However, at least at this point, the fault at the plant site is highly likely an active fault that needs to be considered in terms of seismic design," the report said.

The draft report said the fault or zone of small rocks and sediment called D-1 could move simultaneously with a major active fault called the Urazoko, which is about 200 to 300 meters from the reactor buildings.

If that happens, "it could impact the critical facility right above," the report said.

The report pointed out that the D-1 zone is possibly an extension of the K fault, which showed traces it has moved within the past 130,000 years or so.

It also said the K fault would probably move together with the Urazoko fault.

The safety guidelines for nuclear power plants do not allow utilities to build reactors or critical cooling equipment directly above an active fault. The guidelines define a fault as active if it has moved within the past 120,000 to 130,000 years. The NRA is planning new rules that will expand that time frame by thousands of years.

During the meeting Monday at which the draft report was submitted, Kunihiko Shimazaki, an NRA commissioner who heads the panel, said the report will be checked by other outside experts before it is finalized.

The panel investigated the Tsuruga site in December, when its experts quickly leaned toward the conclusion that the fault appeared active.

Japan Atomic Power has criticized the panel, saying its view lacks scientific support.

NRA chief Shunichi Tanaka has said the Tsuruga plant's No. 2 reactor will not undergo the safety exam required to be restarted if D-1 is active, because the unit will no longer be allowed to operate.

The Tsuruga plant has two reactors, with two faults also running under reactor 1, and Japan Atomic Power is still examining them. But reactor 1 is over 40 years old, and the new rules will not allow reactors over 40 to run in principle. The NRA has been examining possible active faults at other plants, including Kansai Electric Power Co.'s Oi atomic plant, also in Fukui Prefecture.

Fig. 4.4 Newspaper article of *The Japan Times* on January 29, 2013, reporting assessment result of the fault in the Tsuruga Nuclear Power Plant premise by the Nuclear Regulation Authority (NRA). Reprinted by permission from Japan Times

team concluded that the methodology and data supporting point (1) were insufficient, that point (2) was based on an analysis of a very small mineral sample and was thus unreliable, and that the use of minute directionality differences to claim faults that are distinct in point (3) was unreasonable. In addition, the geological horizon of faulting could not be identified. In other words, it could not be concluded that no fault activity had occurred after which formation had accumulated. The aforementioned peer review also confirmed that these judgments are standard in academia. Furthermore, the K Fault is almost certainly active and the presence of a fault of such significance is a grave issue. An attempt to make conservative judgments that take the accuracy and validity of survey results into consideration would necessarily lead to the conclusion that this is structurally relevant to the D-1 Fault.

The fifth assessment meeting held on May 15, 2013, concluded with the final assessment report stating that the assessment results could be reconsidered should significant new information be obtained in the future. The conclusions were approved by the NRA on May 22, 2013 [7].

4.5 Various Criticisms from the Operator

The operator did not accept the NRA's conclusions. They criticized the assessment meeting for being an unscientific gathering that gave preference to opinions from a select handful of experts. They asserted that the NRA should provide concrete evidence if they were to claim that the fault was active. In addition, they asked experts formerly involved in reviews, who had expressed views aligned with their own, to file a document of dissent. On May 21, 2013, they held a press conference at which non-Japanese researchers offered opinions countering the NRA's conclusions. Furthermore, they sent letters of protest addressed to individual members of the survey team [8].

This situation was reminiscent of how safety review frameworks had been conducted in the past. The Fukushima Daiichi nuclear disaster made it clear that a single-operator organization cannot take full responsibility should a nuclear disaster happen. No one will be satisfied with resumption of nuclear power plant operations unless the national government can assure safety or establish disaster measures (including compensation). The NRA's judgments are important precisely for this reason and they should not be influenced by the needs of operators. In addition, the most important aspect is clarity in judgment standards; the NRA cannot do anything but follow the rules if they are to be held responsible for their judgments.

Explanations to citizens would also be meaningless if neutrality were not maintained. Operators still seem to feel no shame in declaring that they will carry out surveys to prove that active faults are absent. The truth, however, had never been made clear before starting the surveys. Furthermore, although it is important to seek third-party opinion, neutrality would not be maintained if the operators themselves selected the experts and offered them compensation.

Operators and a handful of researchers who support them carried out the following three negative campaigns on the current assessment results [9].

The first was the criticism that the expertise of assessment members was unevenly distributed. In addition to Dr. Shimazaki, who is a seismologist, the Tsuruga survey team included one structural geologist and three tectonic geomorphologists. Despite the high representation of tectonic geomorphologists, the expertise of these members collectively covered many important fields. Tectonic geomorphology generally examines the geology that comprises geomorphological features. It also covers a considerable area of Quaternary geology. In the fault survey at Tsuruga, the observation results of its geology were a major topic of discussion. Furthermore, the conclusions of the survey team, based on the results of fracture zone analysis, were supported and reinforced by other expert earthquake geologists and structural geologists during the peer review. Therefore, this criticism completely misses the point.

The second criticism was that the judgment was too conservative (excessive emphasis placed on safety). With regard to nuclear power regulations, Regulation Guide 2006 requires "conservative judgments." However, the operator's criticism was that the identification of an active fault underneath a nuclear reactor constitutes a considerable management problem and that active faults should not be declared unless their presence can be proven with 100% certainty. The views of the management are understandable; however, judgments cannot be considered conservative otherwise. Ensuring there are no active faults underneath important facilities is the responsibility of the operator.

The problem, then, is whether the conservativeness of the judgment was excessive. What the regulation criteria require is a "conservative judgment that takes into account the limitations of science." This means that an objective assessment should be conducted using strict scientific methods and that any uncertainties that remain should not be ignored but assessed conservatively following a certain set of rules. For the Tsuruga Nuclear Power Plant, a fault that was considered extremely likely to be active (K Fault) did in fact exist. In addition, the likelihood that it is part of the same chain of faults as the D-1 Fault, which extends beneath the nuclear reactors, is also high. This explains why it was considered an active fault that should be examined for seismic design. It is an example of precisely the type of essential, conservative judgment required by the regulation criteria.

The third criticism was that the process was not scientific. First, "science" refers to "the search for truth"; it is not about measuring something with equipment and obtaining numbers or solving seemingly sophisticated equations. Obtaining the truth is not an easy task, and simply categorizing things as black and white is not science. Deciding that faults should extend in a straight line or simply considering that two faults are distinct based on only minute differences in directionality goes against the scientific method. In the Tsuruga Nuclear Power Plant survey, we discussed whether it was appropriate for the geological age to be determined from such a tiny sample of volcanic ash. One cannot use this type of uncertain evidence, which would normally be discounted in academia, to claim that a fault is not active because it has not shown activity in the last 120,000 years.

Science is not a tool. It is unforgivable to adopt arbitrary interpretation of the rules to claim that anything is acceptable and to use it whenever convenient. Science was never something that could make clear distinctions between black and white. Making authentic scientific assessments, including acknowledging uncertainty and making judgments that align with the regulatory criteria, is what was required of us.

4.6 Criticisms and Praise from the Media

The Tsuruga Nuclear Power Plant assessment meetings, which spanned almost 6 months, provided a good opportunity for the media and public to deepen their understanding of active faults. Media outlets made the following comments regarding the current assessment.

The *Yomiuri Shimbun* published an editorial in their morning paper on May 16 saying, "The conclusion that the fault was active was reached prematurely," given the objections raised by the operators. They wrote, "… Can they say that the conclusions are scientifically sound enough? This is extremely questionable … scientific assessments should be the fruit of ample data and discussions by a wide variety of experts. The NRA should not make haste to reach a conclusion while Japan Atomic Power Company's investigations are still ongoing." Conversely, the *Nikkei* reported the following in their morning paper on May 17 under the title, "An active fault identification that shows change in nuclear power regulations." They wrote, "[This was] a judgment that demonstrates that nuclear power regulations have taken a significant turn towards placing utmost priority on safety … that the NRA—born out of past lessons—identified active faults based only on scientific evidence and made independent judgment fulfills the purpose of their being … however, there is room for improvement for the NRA as well … with regards to active fault surveys, were the expert selection process and the way in which the discussion proceeded ideal for the purposes of making scientific and highly neutral judgments? … It would not be right to say that the NRA has sufficiently achieved accountability … some nuclear power plants may be forced to terminate operation. However, these plants were constructed in line with national policy and approvals were given by the state. To leave the clean-up to electric power companies alone is an inconsiderate move."

The *Asahi Shimbun* published the following editorial in their morning paper on May 16, under the title, "Tsuruga Nuclear Power Plant—a yellow card was obviously deserved." They wrote, "The current discussion was carried out in a reasonable manner, and the Asahi supports the conclusions … we would especially like to commend that the expert meeting was comprised of researchers who had been uninvolved in nuclear power plant safety reviews and were able to rid past routine … this was essentially an attempt to enforce safety regulations from an independent standpoint. This stance should be firmly maintained in future assessments … in order to give top priority to decisions made from the latest scientific findings, the state may consider offering some form of support for operators, who will suffer significant loss." They also expressed concern regarding certain moves in politics and stated, "Gradually retreating to 'pre-3/11' conditions would be impermissible."

The *Mainichi Shimbun* published the following editorial in their morning paper on May 17 under the title "Tsuruga Nuclear Power Plant Unit 2—urgent need to prepare for decommissioning." They wrote, "… From the standpoint of ensuring safety in nuclear power plants, the Mainichi Shimbun supports the conclusions presented by the survey team… Japan Atomic Power Company claims that the judgment was impractical. However, the NRA made a reasonable decision by classifying faults as active unless the possibility could be rejected … there will be a major impact on the finances of local authorities and the local economy, which have depended on nuclear power plant sites. The state, which approved the construction of Unit 2 in Tsuruga, is also responsible. Preparations for decommissioning, such as discussions on how expenses for decommissioning should be funded and new plans for regional development, must be carried out …. Decommissioning plans for Tsuruga's Unit 2 will be a touchstone for the new, nuclear-free era to come."

The *Tokyo Shimbun* (and *Chunichi Shimbun*) published the following editorial in their morning paper on May 23 under the title, "The NRA identifies active fault—a step towards a country that prioritizes safety." They wrote, "The NRA's decision to classify a fault underneath nuclear power plants as active is a significant turning point in the history of nuclear energy in Japan. This is a new step towards prioritizing safety over economy … This finally makes the myth of safety in nuclear power plants a thing of the past and brings hope for a future with a culture of safety and a science of safety …. Massive earthquakes can occur at any time. Prolonged surveys increase risk with every additional day. The Tokyo Shimbun supports the NRA's decision, which prioritized safety. Most citizens would also likely support their conclusions … with regards to earthquakes, we hope (the NRA) will maintain their stance of making conservative judgments on uncertain issues. Nuclear power plants should self-initiate reexaminations for purposes of prioritizing safety … the state should not hold back from giving support to mend mistakes in (former) national policy."

On the May 16 airing of NHK's Jiji Koron, commentator Noriyuki Mizuno expressed the following under the title "The active fault problem in nuclear power plants—sorting out issues toward decommissioning." He said, "Regulations before the Fukushima Daiichi disaster were criticized for obliging to electric power companies' needs. The NRA's judgment is a clear step away from former regulatory frameworks, as it placed utmost priority on safety security. They deserve appraisal for this … until now, neither the state nor electric power companies accepted that faults were active unless there was concrete evidence, nor did they prepare countermeasures … electric power companies' in-premise surveys were insufficient. They lacked the sincerity to respond to concerns raised by experts. The fact that the state left surveys to electric power companies and ended up overlooking active faults is a problem … if we take into account lessons learned from the disaster, regulatory organizations will need to make conservative judgments for matters that potentially pose a significant threat to citizens … the NRA should decide not to approve resumption of operation … the Abe administration … plans to put nuclear power plants back into service once their safety is confirmed. If so, it is only sensible that nuclear power plants which the NRA deems unsafe be instructed to terminate service. The state is also accountable for having overlooked the possibility of active faults. The government should urgently discuss challenges with decommissioning and outline a framework that citizens can accept" (http://www.nhk.or.jp/kaisetsu-blog/100/).

In addition, *The Japan Times* published the following editorial in their paper on May 20, 2013, under the title, "Tsuruga reactor's active fault." They wrote, "A five-member panel of the Nuclear Regulation Authority last Wednesday issued a report saying that the D-1 fracture zone of pebbles and sediment running beneath the No. 2 reactor of Japan Atomic Power Co's Tsuruga nuclear power plant in Fukui Prefecture is an active fault. Because constructing important nuclear-related facilities above an active fault is banned, the company will very likely have no other choice than to decommission the reactor. On the basis of the report, the NRA is expected to decide soon not to allow the restart of the reactor. The NRA should not bow to the pressure that is expected to come from the company and the Liberal Democratic Party. It should make a decision solely relying on convincing scientific findings … The panel held its first meeting in October 2012. Japan Atomic Power was allowed to submit its own data. In March 2013 the panel held a peer review to listen to opinions from experts who did not take part in the on-the-spot geological study of the nuclear power plant site. It took a fairly long time to issue the report because of the strong resistance put up by Japan Atomic Power. But it must be remembered that experts expressed a view by the early 1990s that the Urazoko fault is active. Even the now-defunct, infamous Nuclear and Industrial Safety Agency in 2011 expressed suspicions that the D-1 fracture zone is an active fault. NISA also expressed a suspicion that active faults existed in the compounds of five more nuclear power stations. The NRA should stick to the principle that if the power companies concerned cannot convincingly prove that the faults in question are not active, it will not allow the operation of the nuclear power plants … The government should consider ways to economically diversify communities hosting nuclear power plants that are facing closure to make the transition easier for residents."

References

1. Tsutsumi H, Toda S (2012) Surface rupture associated with the April 11, 2011 Fukushima-ken Hamadori earthquake and Paleoseismic history of the causative fault. J Geol Soc Jpn 118:559–570. (in Japanese)
2. Suzuki Y, Watanabe M (2012) Problem on active fault assessment in the Ooi nuclear power plant and necessary countermeasures for safety regulation. Kagaku 82:854–856. (in Japanese)
3. Asahi Shimbun (2012) Starting fault investigation in Oi Nuclear Power Plant premise. Evening Paper on 2nd Nov 2012. (in Japanese)
4. Japan Atomic Power Company (1965) License application for Unit 1 of Tsuruga nuclear power plant. (in Japanese)

5. Suzuki Y (2013) Significance and problem on fault survey in the Tsuruga nuclear power plant by Nuclear Regulation Authority of Japan. Kagaku 83:111–112. (in Japanese)
6. Suzuki Y (2013) Nuclear power plants and active faults – “unanticipated” is unacceptable. Iwanami Shoten, Tokyo (in Japanese)
7. Nuclear Regulation Authority (2013) Assessment of crush zone in Tsuruga Nuclear Power Plant premise. http://www.nsr.go.jp/committee/kisei/data/0007_02.pdf. (in Japanese)
8. Tokyo Shimbun (2013) Sending letters of protest addressed to individual members of the survey team, Morning Paper on 22nd May, 2013. (in Japanese)
9. Watanabe M (2013) Active fault investigation by Nuclear Regulation Authority. Kagaku 83:149–152. (in Japanese)

Chapter 5
How Should We Move Forward? New Regulatory Requirement and Frameworks

Abstract In order not to repeat the consequences of underestimation of active faults, it is necessary to establish clear rules and tighten operations. The NRA finalized the New Regulatory Requirements for Light-Water Nuclear Power Plants ("New Regulatory Requirements") on June 19, 2013. The fulfillment of these requirements is required by law, and the existing facilities must meet the requirements in order for them to continue operation. The safety of the nuclear power plants has been re-reviewed following these criteria from July 2013 onward. The author has participated in the "Regulation Standard Review Team for Earthquakes and Tsunamis" since November 2012, reviewed the existing rules, and drafted the standards for regulations. The contents of the new standards basically followed the conventional rules, and strict operation was the primary area of focus. The establishment of regulatory standards can be expected to greatly improve nuclear safety reviews. However, a fair research system that includes third parties and strengthening the system of the NRA remain as future challenges. It is also of paramount importance to take the limitations of science seriously and make appropriate social judgments without overconfidence.

Keywords New regulatory requirements · Safety regulations · Scientific judgment · Overconfidence

5.1 Strict Rule Enforcement Based on New Regulatory Requirements

There is nothing more troublesome than finding oversight in active faults after nuclear power plants have been built. Translocation would be difficult if the plant was built above a fault and, in addition, the seismic design might need to be completely reconsidered. Although scientific limitations pose challenges in making

Y. Suzuki, *Active Faults and Nuclear Regulation*,
https://doi.org/10.1007/978-981-15-0765-6_5

assessment results that are completely correct, this is precisely the reason why it is important to make judgments through strict regulatory adherence to achieve accountability. This is relevant to whether everyone will be satisfied when the unexpected occurs. We should learn from the fact that confusion lingers in the present day since this was not realized in the past.

The NRA finalized the New Regulatory Requirements for Light-Water Nuclear Power Plants ("New Regulatory Requirements") on June 19, 2013. According to the NRA, major safety regulation problems before the Fukushima Daiichi nuclear disaster were (1) regulatory requirements that did not cover "severe accidents" and few preventive activities in place and (2) the lack of a legal framework to retroactively apply new requirements to existing nuclear power plants, which hindered continuous safety improvements. Therefore, the New Regulatory Requirements have strong legal power, and existing facilities must meet the requirements in order for them to continue operation. Nuclear power plant safety is to be re-reviewed following these criteria from July 2013 onward. Of essential importance is that the thinking underlying the new criteria be publicized (http://www.nsr.go.jp/english/regulatory/index.html).

Reinforcement of the requirements was necessary to ensure that nuclear disasters never happen again because of a once in a millennium event such as the 2011 Great East Japan Earthquake and tsunami. Although some say the tsunami was the sole cause of the Fukushima Daiichi nuclear disaster, a review team on earthquake and tsunami regulatory requirements was established to discuss how the threat of low-frequency huge earthquakes should be addressed, including the countermeasures against seismic shaking. They concluded that extremely rare worst-case scenarios must never be ignored [1].

The New Regulatory Requirements take into account the scientific limitations regarding seismic shaking and fault activity prediction. They clearly show the stance that sites at which consequences cannot be predicted adequately, e.g., on top of faults, must be avoided. They prohibited the construction of important facilities above active faults and they enforced the same for sites above landslides that rupture foundation rocks. The traditional opinion that construction above faults should be allowed as long as the facility is considered structurally tolerant was widely supported. However, the new criteria do not adopt this position because current science cannot guarantee safety with sufficient reliability. In the past, it was almost as if constructing nuclear power plants in less-than-ideal sites and complementing that with technology was a national policy. In contrast, since the 2011 Great East Japan Earthquake, we appear to have adopted a new approach toward design that recognizes construction should not be forced, especially nuclear power plants [2].

Strict enforcement was also attempted for active fault identification. As mentioned before, Revised Regulatory Guide 2006 defined active faults that should be examined for seismic design as "faults in which activity after the late Pleistocene cannot be eliminated." It also outlined that faults should be considered active if the possibility of activity cannot be ruled out. This, however, had not been strictly enforced and it had invited confusion over active fault identification. Although concrete evidence of inactivity after the late Pleistocene must be presented in order to classify faults as inactive, activity had been denied simply by presenting the lack of evidence of activity.

One suggested solution for deterring such inappropriate identification was to prevent oversight by widening the applicable target range, in particular, by unifying the definition of active faults to the standard scientific understanding that they are "faults that showed activity in the last 400,000 years." The underlying reasons were that faults that have moved in the last 400,000 years are typically active from the perspective of active tectonics and their recurrence intervals are unlikely to exceed 100,000 years. Therefore, the lack of evident activity in the previous 120,000 years might stem simply from the technical difficulty in detecting it.

Ultimately, however, the clause defining active faults that should be examined for seismic design was not revised. The reasoning was that the problem lay not with the definition of active faults but with adherence to the rules. The situation will not improve if the rules are administered in the same manner, even if the definition is revised, and any major change in definition is susceptible to opposition. The "Detailed Note on Requirements" outlines that (1) surveys must cover at least the previous 400,000 years, and (2) if activity appears to have occurred in the previous 400,000 years, the possibility of the fault being active should be deliberated to avoid misjudgment.

To incorporate the lessons learned from the past and to prevent problems regarding the definition of active faults, the use of the term "active fault" was discontinued and replaced with the unifying expression, "faults etc. that might exhibit activity in the future." The concept of the revised wording is almost the same as "active faults that should be examined for seismic design," mentioned in Revised Regulatory Guide 2006. However, the new expression strictly eliminates the possibility for narrow interpretation, i.e., active faults only include major faults capable of causing earthquakes.[1] Regardless of length or depth, "faults etc. that might exhibit activity in the future" would cause serious damage once activated, if located inside nuclear power plant sites. The same could be said of normal faults or landslide slips. This falls under the category of "capable faults," according to the IAEA rules. Clear stipulation of the above led Japan to match international standards.[2]

[1] In order to decide on appropriate countermeasures, prior to the New Regulatory Requirements, the idea has been proposed to distinguish active faults into several categories: (1) causative faults, (2) secondary faults, and (3) other faults. Causative faults are faults that could be the origin of an earthquake. Secondary faults have some structural relationship with causative faults and move in association with them. Other faults are those elicited by shaking, land deformation, etc. The former Nuclear Safety Commission stated that construction of important facilities above (1) and (2) should effectively be prohibited, but that this would not apply to (3). Although this seems logical, specific categorization is difficult, especially for (2) and (3), and it is largely influenced by individual interpretation. It is possible that what used to be considered (3) could become (2), if (1) was found nearby. This is one reason why this concept was not incorporated in the current regulations.

[2] The IAEA's "Seismic Hazards in Site Evaluation for Nuclear Installations, Specific Safety Guide, No. SSG-9, Vienna, 2010" stipulates the following. "Fault displacement can occur as a result of an earthquake (either directly or indirectly). In other words, displacements could be associated

Although NRA Chairperson Dr. Shunichi Tanaka praised the New Regulatory Requirements as being the strictest, this should not be surprising given the occurrence of the Fukushima Daiichi nuclear disaster. I was once caught speechless when I heard the opinion that "Japan should not make the rules too strict, since earthquakes are frequent - otherwise nuclear power plants could not be built." This is utter nonsense and I hope that this never happens again.

Using uncertain evidence that would not be approved in academic standards to deny the possibility of future activity, as was the case with the onsite fault surveys at the Tsuruga Nuclear Power Plant, should not be tolerated. It must be mandatory to consider scientific limitations and to make conservative judgments that acknowledge uncertainty. Nuclear plant operators raised objections to the new requirements, claiming them to be excessively strict. However, their content was the result of honest acceptance of the science and technology standards of the present day. Only by meeting strict regulatory requirements can we pave the way to recover trust in nuclear power plant operations.

The most important thing is strict adherence to the New Regulatory Requirements. Instead of being fixated on nuanced interpretations of expressions used in the requirements, we must maintain sincerity toward the science and the spirit of prioritizing safety. Judgments on whether requirements are met should be left to the discretion of the NRA to avoid unnecessary prolonged argument. For many years, we were aware that recurrence of the Jogan Earthquake could initiate a giant tsunami. We should take to heart that, regardless of what we knew, we were unable to act immediately and that it was all too late.

The Japan Times published the following editorial in their paper on June 28, 2013, under the title "New nuclear safety standards." They wrote, "The Nuclear Regulation Authority on June 19 approved new safety standards for reactors. It will put them into force beginning July8. … NRA Chairman Shunichi Tanaka said that his organization will not take power companies' business conditions into consideration. The NRA should stick to this principle and carry out the screening in a transparent manner and reject any political pressure. Although the Abe administration and the power industry are eager to restart nuclear power plants, they should not view the new safety standards as a procedural tool for the restart. Power companies must understand that if reactors fail to meet the requirements under the new standards, the companies must start a concrete process of decommissioning them. The essence of the new standards is that they require power companies to take measures to cope with a severe accident as happened at Tokyo Electric Power Co.'s Fukushima No. 1 nuclear power plant. … Before the Fukushima nuclear fiasco, operators of nuclear power plants were required to find out whether geological faults inside a

---

with a causative fault or could occur co-seismically on secondary faults. It should be noted that tectonic displacements associated with folds (synclines and anticlines) are also included in the term 'fault displacement' (https://www-pub.iaea.org/MTCD/Publications/PDF/Pub1448_web.pdf)." In terms of measures, both consideration of alternative sites for new sites and detailed examinations of existing sites are required; however, instructions regarding judgments are not clearly outlined.

plan site were active in the past 120,000 to 130,000 years. The period to be examined will now be extended to the past 400,000 years if the operators are unable to prove that faults have not been active in the past 120,000 to 130,000 years. Important facilities of a nuclear power plant must not be built near an active fault. The new standards are based on the idea of backfitting: making changes to existing nuclear power plants to reflect the latest knowledge and findings related to safety. The NRA should quickly incorporate such knowledge and findings into the standards …" (Fig. 5.1).

5.2 Reinforcing Nuclear Regulation Authority Frameworks and Securing Trust

In the past, the entire process from survey to assessment was conducted by the operators and the regulatory authorities only reviewed its validity. From now on, however, the NRA has a public mandate to judge safety. Judgments are left to the discretion of the NRA and operators must present appropriate documentation to allow the NRA to fulfill this purpose. The New Regulatory Requirements or Detailed Note on Requirements should not capture the regulating authorities; instead, depending on the latest findings, they should of course be allowed to apply strict judgments that exceed what is outlined in the requirements.

EDITORIAL

New nuclear safety standards

The Nuclear Regulation Authority on June 19 approved new safety standards for reactors. It will put them into force beginning July 8. Four power companies are expected to apply for safety screening by the NRA in July to restart up to 12 reactors at six nuclear power plants.

NRA Chairman Shunichi Tanaka said that his organization will not take power companies' business conditions into consideration. The NRA should stick to this principle and carry out the screening in a transparent manner and reject any political pressure.

Although the Abe administration and the power industry are eager to restart nuclear power plants, they should not view the new safety standards as a procedural tool for the restart. Power companies must understand that if reactors fail to meet the requirements under the new standards, the companies must start a concrete process of decommissioning them.

The essence of the new standards is that they require power companies to take measures to cope with a severe accident as happened at Tokyo Electric Power Co.'s Fukushima No. 1 nuclear power plant. Such measures include construction of sea walls that will protect individual nuclear power plants from the largest tsunami that could conceivably hit them and installation of filters to remove radioactive substances if such substances are vented from reactor cores into the atmosphere during an emergency.

Before the Fukushima nuclear crisis, authorities had left the decision on whether to adopt those measures up to power companies by virtually refusing to consider the possibility that a severe accident could happen. But under the new safety standards, a grace period of five years will be allowed for the installation of filters in pressurized light-water reactors, since hydrogen explosions are unlikely in them. There is no grace period for installing filters in boiling-water reactors.

A grace period of the same length will be allowed for the installation of a second control room that is capable of independently controlling and cooling reactors in the event of a natural disaster or a terrorist attack.

Before the Fukushima nuclear fiasco, operators of nuclear power plants were required to find out whether geological faults inside a plan site were active in the past 120,000 to 130,000 years. The period to be examined will now be extended to the past 400,000 years if the operators are unable to prove that faults have not been active in the past 120,000 to 130,000 years. Important facilities of a nuclear power plant must not be built near an active fault.

The new standards are based on the idea of backfitting: making changes to existing nuclear power plants to reflect the latest knowledge and findings related to safety. The NRA should quickly incorporate such knowledge and findings into the standards.

The government, for its part, should adopt a policy to speedily reduce Japan's reliance on nuclear power and to eventually abolish it. It also should help create new industries in areas now hosting nuclear power plants.

Fig. 5.1 Editorials of *The Japan Times* on June 28, 2013, reporting on the importance of New Regulatory Requirements approved by NRA. Reprinted by permission from *The Japan Times*

The role of the NRA will become increasingly important. In order to protect public safety, the NRA is responsible for reviewing safety in individual nuclear power plants and for selecting those plants that could be reintroduced to service. As judgments regarding safety relate directly to the management of the electric power companies, the NRA could face increasing pressure from other sources. However, strict judgment regarding safety is paramount; therefore, the review must guarantee a high level of expertise, a solid neutral stance, and unbiased transparency.

Active fault surveys tend to be dependent on results obtained from a select area of the target region, given the limitation that the entire target area cannot be excavated. Consequently, survey plans determine success. Operators do not employ experts on earthquakes or active faults nor do they show enthusiasm for finding active faults. Under these circumstances, it is fair to assert that effective safety reviews could not be conducted if the entire process from survey planning to execution were left to the operators.

Ideally, the NRA or a completely neutral third party should conduct the surveys. However, in addition to the fact that the organizational framework of the NRA Secretariat is insufficient, there are discussions about whether the state should cover survey expenses for privately run businesses and concern about whether a vendor able to undertake strict reviews could be secured. Many experts have expressed the opinion that immediate resolution would be difficult.

I believe the NRA Secretariat should urgently achieve independence in expertise so that they can conduct their own active fault surveys. If designing such a system requires time, leaving surveys to the operators during that period might be unavoidable. In that case, I would like to propose the following two points that are achievable immediately.

First, the NRA should take the lead in planning surveys, which are fundamental to determining whether active faults can be discovered. It is clear that operators are unenthusiastic about finding active faults. At this rate, the NRA will be unable to achieve full accountability regarding final assessment conclusions.

Second, it should be made clear that surveys conducted by the operator are carried out on behalf of the NRA (although expense burdens would remain the same). As judgments are made by the NRA and not by the operators, staff from the NRA Secretariat must observe the situation periodically, report to the NRA, and instruct operators as required.

Still in its early stage, the NRA's stance on safety regulations is clear and it appears to be meeting the expectations of much of the public. Conversely, operators and local communities have been left in some confusion because what nuclear power plant administrators previously considered acceptable and approved has suddenly been overturned. It is clear that we must apply lessons learned from the 2011 Great East Japan Earthquake and the Fukushima Daiichi nuclear disaster to change direction regarding nuclear power regulation. The confusion of the present day is an issue that should be resolved by Japanese society as a whole.

5.3 Toward Establishing the Fundamentals of Scientific Safety Judgment

The active fault problem tends to remain blurred because activity frequency is low. Current actions by the NRA are based on serious reflection over how prioritization of economic factors and lack of anticipation of huge tsunamis that occur once in every 500–1000 years led to the disaster at the Fukushima Daiichi Nuclear Power Plant. Therefore, it was only natural that strict measures be taken to deal with active faults. A fundamental shift in the course of nuclear safety is necessary to remedy the lack of anticipation and now is the optimum time for its realization. It would be ideal to reaffirm that threats from active faults could be averted by careful measure setting and to keep the discussion that "(active faults) should not have to be examined to such an extensive degree" from resurfacing again, bearing in mind the characteristics of nuclear power plants.

Some people say that active faults are not dangerous because they rarely move. Others claim onsite faults represent only a minor issue because they are smaller in scale in comparison with typical faults. People might dismiss the risk associated with active faults because of their low frequency of occurrence; however, under-evaluation of active faults at nuclear power plant sites must not be tolerated. I hope these opinions do not become perceived in a distorted way that could invite negligence.

Probability theories can also add to the complication of the situation. Determining the probability of risks that should be considered is important for an objective grasp of the situation. However, I cannot agree with the approach that threats that occur less than once every 10,000 years should be completely discounted. Some say "… there is no such thing as definite safety. The likelihood of a meteor falling and destroying Japan is not zero" and "… so worrying about active faults is pointless." This is clearly a ridiculous standpoint. We cannot prepare for meteors because there is no way of telling where they might fall. That is not the case with active faults. If we know where active faults are, we can avoid building on those sites; if we find out after construction, we can decide to decommission. If we did not do as such, we would be unable to come to terms with the situation in the event that disaster happens. In such a circumstance, responsibility will obviously be questioned.

Whether a site should be a target of disaster prevention measures should not be based on how large the probability is but on whether it is acceptable to the public. Thus, decisions should not be made single-handedly by experts but with public approval. There should be distinction between matters that do or do not have "predictability" and "avoidability." We should not simply consider probability numbers, but instead reflect on what they represent. Even if we do not know when they might move, adopting appropriate measures in relation to known active faults should be a basic rule, at least for nuclear power plants. We should take the opportunity to recognize this as a form of disaster prevention and to come to accept it [3]. To avoid repeating the same regretful mistake of the lack of anticipation highlighted by the 2011 Great East Japan Earthquake, we should accept "science" with sincerity and strive to make appropriate social judgments.

References

1. Suzuki Y (2013) In order never to repeat nuclear accident. Kagaku 83:241. (in Japanese)
2. Nuclear Regulation Authority (2012) Minutes of Subcommittee Meeting on Guidelines for reviewing seismic safety concerning geological faults. http://www.nsr.go.jp/committee/yuushikisya/shin_taishinkijyun/. (in Japanese)
3. Suzuki Y (2012) Active fault in Japan—geographical description and theoretical consideration on its disaster. Gakushikai-kaiho 896:90–94. (in Japanese)

Afterword

Sometimes I hear people say that "The nuclear power plant problem should be discussed holistically bearing in mind future energy problems. Preaching about the earthquake/tsunami problem extensively, or worse yet, emphasizing active faults that rarely ever cause earthquakes is concerning." Balance is indeed important; however, I believe that holistic and rightful decisions cannot be made unless important points of discussion are addressed individually. This is why I knowingly focused on active faults without fear of criticism.

In addition to conducting basic research on active faults, I have studied disaster prevention regarding active faults, which became a social issue following the Kobe Earthquake in 1995. Since 2005, I have also examined records and documents from nuclear power plant reviews, analyzed aerial photos, and undertaken field surveys. I came to realize that active faults near nuclear power plants had been underrated in many cases. I then began questioning whether it was permissible for me, as an active fault researcher, to be ignorant of the nuclear power plant problem. Moreover, I became aware that the field of active fault research had been denigrated, with people saying, "active fault identification is subjective nonsense." It was frustrating for me to realize that the field was being undervalued [1].

I have made sharp criticisms regarding the sloppiness of active fault assessment and the inadequacy of review frameworks in the field of nuclear regulation. It was my hope to resolve the issues from the past and to somehow renew the framework. To achieve this, human failings could not be ignored, which led to criticisms of people with whom I shared the same goals [2, 3]. This heartbreaking move destroyed personal relationships within the academic community. However, my judgment was that the problems were something I could see perceive precisely because they were within my field of expertise; thus, I had no choice but to state my case in order to renew the framework.

I am not optimistic about nuclear energy by any means because it continues to have various issues. My opinion, however, was that continued operation of nuclear power plants for a certain period was unavoidable, because setting a new direction for resolving Japan's energy issues would take time. In this circumstance, I believed

Y. Suzuki, *Active Faults and Nuclear Regulation*,
https://doi.org/10.1007/978-981-15-0765-6

that securing the safety of nuclear power plants, as much as possible, was highly important.

It was within this context that the nuclear disaster occurred following the 2011 Great East Japan Earthquake. The accident forced me to question the mission of active fault researchers in academia, and I have yet to reach a clear conclusion. I do believe it is important to refine academic advice in order to transition from a discussion focused on economics to one focused on safety, and to be mentally prepared to lead society in that direction. The idea that economic activity is precisely the thing that brings social happiness is understandable. However, this idea can be tainted by vested interests and inconsiderate of human rights and ethical issues. In addition, we must continue to check that decisions are responsible and not short sighted with consideration only for immediate gains. Above all else, I believe our mission is to ensure science is not used for convenience, and that it continues to contribute to the social good.

I was surprised to see that nuclear plant operators were openly critical of the NRA for its active fault assessment. I especially felt concern over the misguided way in which they used the word "science." Seeing this renewed my determination to transform the spirit and framework under which previous reviews had been conducted, and to start things fresh and anew.

The NRA will soon start considering those nuclear power plants that may resume service. Safety security must be paramount for re-servicing nuclear power plants and strict reviews are desirable. Clearly, the question of whether the use of a nuclear power plant is acceptable and problems associated with nuclear power plant safety are two distinct issues that should be discussed separately. Safety must not be compromised; accepting the possibility of a grave disaster simply because the occurrence probability is low is impermissible. The last thing I wish is that we are again faced with regret should the unanticipated happen.

References

1. Watanabe M (2009) Problem in the examination of active faults by nuclear regulation authority. Kagaku **79**:179–181. (in Japanese)
2. Nakata T (2011) Inaccurate assessment on active fault in nuclear regulation. Asahi J **116**(25):33–35. (in Japanese)
3. Watanabe M, Nakata T, Suzuki Y (2012) Active faults and nuclear power plants - who is falsifying the regulation? Sekai **825**:125–134. (in Japanese)

Appendix A: Summary of "Outline of New Regulatory Requirements for Light Water Nuclear Power Plants, Regarding Earthquakes and Tsunamis (Provisional Translation)" Publicized by NRA Website: http://www.nsr.go.jp/english/regulatory/index.html

I Basic Design Policy for Earthquakes and Tsunamis

[Basic requirements] <Standard for Establishment License>

1. *Since a nuclear power reactor facility (hereafter called "facility") must be provided with high safety as a whole, the following basic design policy shall be satisfied.*
   1.1. *Facilities with important safety functions shall be established on the ground that has been confirmed to have no outcrop of a fault, etc. with the possibility of becoming active in the future.*
   1.2. *Facilities with important safety functions shall be designed in such a way that it does not lose these important functions against a seismic force caused by the earthquake ground motion (hereafter called "standard seismic motion") that rarely occurs, but may possibly occur in service period of the facilities and have a significant effect on the facility. In addition, the facility shall be designed in such a way that it can fully sustain a seismic force assumed to be appropriate in accordance with importance in view of the effect of radiation on the environment following a loss of safety functions that may possibly occur due to an earthquake.*
   1.3. *Facilities shall be established on the ground with sufficient support capacity against the seismic force provided above.*
   1.4. Facilities with important safety functions shall be designed in such a way that it does not lose its safety functions against a tsunami (hereafter called "design basis tsunami") that rarely occurs, but may possibly occur in service period of the facilities and have a significant effect on the facility.
2. With regards to survey for formulating the standard seismic motion and design basis tsunami, etc., survey method shall be selected in accordance with the purpose and the reliability and precision of the survey results shall be ensured by

Y. Suzuki, *Active Faults and Nuclear Regulation*,
https://doi.org/10.1007/978-981-15-0765-6

taking into account the application conditions and precision of the survey method.

[Detailed note on requirements] <Internal Rules of the NRA>

(1) *A “fault, etc. with the possibility of becoming active in the future” includes a landslide surface that cuts the bearing basement in addition to an active fault taken into account as the hypocenter and a fault that is accompanied by the activity of the fault concerned resulting in a permanent displacement.*
(2) *“fault, etc. with the possibility of becoming active in the future” means fault of which activities later than the Late Pleistocene (later than 120–130,000 years ago) cannot be denied. To judge this, in case activities later than the Late Pleistocene is not clear according to, for instance, lack of plural or continuous strata of Late Pleistocene, its activities further to be investigated for the Middle Pleistocene (later than 400,000 years ago) through the integrated consideration of geography, geology, geological structure and stress field etc. Note that if confirmation is difficult at installation area where activity is assessed, it should be judged to the conservative side by the behavior of fault etc. confirmed at the extended part of the fault concerned.*
(3) *The description of “in particular” in No. 1 of Paragraph 1 is provided because if a facility with important safety functions is established in a place with the outcrop of a fault, etc. with the possibility of future activity, the activities of such a fault, etc. in the future may have a significant effect on the safety functions.*

(Reference)
With regards to the existence of a “residual risks” related to formulating the standard seismic motion and design basis tsunami

It cannot be denied that an earthquake motion occurs with the strength exceeding that of the standard seismic motion in view of seismology. In addition, it cannot be denied that a tsunami exceeding the design basis tsunami will possibly attack facilities. In formulating the standard seismic motion or design basis tsunami, these matters mean the existence of a “residual risks” (a risk that causes a significant damage event to occur in facilities due to the effect exceeding the standard seismic motion or design basis tsunami that has been formulated on them, that an event that diffuses a large amount of radioactive materials from facilities to occur or at these events result in general public to suffer from radiation exposure). Thus, when a facility is designed, the possibility of the occurrence of an event exceeding the formulated standard seismic motion or design basis tsunami should properly be taken into account and the existence of this “residual risks” shall fully be recognized including the stage of the basic design and its subsequent stages to make effort to minimize the risk as far as this approach is feasible.

<The following are major references.>

- Regulatory Guide for Reviewing Seismic Design of Nuclear Power Reactor Facilities (revision draft) (March 22, 2012)
- Guide for Seismic Safety Review of Nuclear Power Reactor Facilities (revision draft) (March 22, 2012)

II Classification of Importance of Facilities

[Basic requirements]

1. Since a facility must be provided with high safety as a whole, all facilities with important safety function shall belong to Class S in view of preventing a loss of safety functions that may possibly occur by an earthquake followed by the effect of radiation on the environment. Those with less effect than that of Class S can be assigned to Class B. Those other than them and with safety equal to that of general industrial facilities and public facilities can be assigned to Class C.
2. Facilities/SSCs which protect Class S facilities provided in item 1 against tsunami effect (hereafter "tsunami protecting facilities/SSCs"), and SSCs which has tsunami monitoring function at the site (hereafter "tsunami monitoring SSCs") shall belong to Class S in view of the effect of a loss of the functions concerned that may possibly occur due to an earthquake.

   [Detailed note on requirements]: omitted

III Formulation of Standard Seismic Motion

[Basic requirements] <Standard for Establishment License>

It is appropriate to formulate the standard seismic motion used for the seismic design of facilities from the seismological and earthquake engineering point of view on geology, geological structure, soil structure, seismicity, etc. of the site and its surrounding area based on the latest scientific and technological knowledge and findings. The standard seismic motion shall be formulated according to the following policies.

1. The standard seismic motion shall be formulated as the earthquake ground motion in the horizontal and vertical directions on the free surface of the base stratum in the site as to the "earthquake ground motion formulated with a hypocenter specified for each site and formulated" in the following paragraph and the "seismic motion formulated without a hypocenter" specified in Paragraph 3, respectively.
2. The "earthquake ground motion formulated with a hypocenter specified for each site" shall be formulated by selecting multiple earthquakes that are predicted to have a significant effect on the site (hereafter called "earthquakes for investigation") as to inland crustal earthquakes, interplate earthquakes and oceanic intraplate earthquakes and implementing the evaluation of ground motions for each selected earthquake for investigation by the method based on response spectra and the method using the fault model, *with uncertainties taken into account* and propagation properties up to the free surface of the base stratum reflected.
3. The "earthquake ground motion formulated without a hypocenter specified" shall be formulated by collecting observation records in the vicinity of a

hypocenter that were obtained as to inland crustal earthquakes difficult to associate a seismic source with an active fault and setting up a response spectrum in accordance with the soil properties of the site with various uncertainties taken into account.

[Detailed note on requirements] <Internal Rules of the NRA>

(1) With regards to the character of the standard seismic motion.
The standard seismic motion is the one that is a prerequisite of the seismic design to ensure the seismic safety of a facility. When it is formulated, its adequacy shall be fully confirmed by checking the latest knowledge and findings at the time of individual application.
(2) The interpretation of terms used when the standard seismic motion is formulated is as follows:

a. The "free surface of the base stratum" refers to a free surface virtually set up assuming that neither surface layer nor structure exists on the base stratum in order to formulate the standard seismic motion and that is the surface of a bed rock postulated with a almost horizontal considerable extent and without remarkable elevation differences. The "base stratum" referred to here is a hard ground with a shear wave velocity of almost Vs = 700 m/s or more without any remarkable weathering.
b. An "inland crustal earthquake" refers to the one that occurs in the seismogenic layer of the upper crust in the continental plate and that includes those occurring slightly offshore from the seashore.
c. An "interplate earthquake" refers to the one that occurs on the boundary between two adjacent plates.
d. An "oceanic intraplate earthquake" refers to the one that occurs inside the subducting ocean plate and it can be divided into two types: the one is the "subducting oceanic intraplate earthquake" that occurs near or slightly seaward of the trench and the other is the "subducted oceanic intraplate earthquake (intraslab earthquake)" that occurs on the landward side of the trench.

(3) An "earthquake ground motion that is formulated with a hypocenter specified for each site" shall be formulated in accordance with the following policies.

a. With regards to an inland crustal earthquake, an interplate earthquake and an ocean intraplate earthquake, the nature of the active fault and the situation of the occurrence of an earthquake shall carefully be examined and existing research results and so on shall comprehensively be studied in relation to the distribution of middle, small and micro types of earthquakes, stress fields and mechanisms (including plate geometry, movement, and interactions.) to select multiple earthquakes (earthquakes for investigation).
b. With regards to inland crustal earthquakes, the following shall be taken into account.

i) *When an active fault taken into account for seismic design is assessed, the properties of existing literature investigations, geomorphological surveys, geological surveys, geophysical survey, etc. shall effectively be utilized and proper combinations of these surveys shall be implemented, Then, their results shall comprehensively be assessed to clarify the location, geometry and activity of the active fault.*

ii) *When the fault geometry and parameters of source characterization are estimated, an isolated and short active fault shall be carefully handled, and simultaneous rupture among multiple active faults shall be taken into account.*

c. With regards to an interplate earthquake and an oceanic intraplate earthquake, their source areas shall be estimated in light of the large-scale earthquakes that occurred both in Japan and the world as well as the similarity of their seismogenic mechanisms and tectonics backgrounds.

d. An earthquake ground motion shall be evaluated for each earthquake for investigation based on the response spectrum method shown in i) as below and by the method using the fault model in ii) and then formulated. Ground motion evaluation should be conducted considering fully on earthquake types, characteristics of seismic wave propagation path (including site effect) etc, based on seismic ground motion observation records at the site.

i) The evaluation of an earthquake ground motion based on the response spectrum method
Response spectra shall be evaluated for each earthquake for investigation by using a proper method and the design basis response spectrum shall be estimated based on these response spectra, while the earthquake ground motion shall be evaluated by properly taking into account its properties such as its duration, time history of the amplitude envelope, etc. based on the scale of an earthquake, hypocentral distance, etc.

ii) Evaluation of the seismic hazard based on the method using the fault model
The ground motions shall be evaluated with determination of the parameters of source characterization by using an appropriate method for each earthquake for investigation.

e. With regards to various uncertainties (the length of a causative fault, the depth to the top and bottom of the seismogenic layer, the dip of a fault, the location and size of an asperity, the stress drop, the rupture starting point, etc. and uncertainties due to a difference in interpretation) accompanying the process of formulation of the standard seismic motion, dominant parameters shall be analyzed that can be considered to have a significant effect on the evaluation of the seismic hazard and they shall be taken into account by using an appropriate method such as combination of uncertainties.

f. *If a hypocenter is extremely close to the site among earthquakes for investigation that were selected in relation to earthquakes in inland crusts, the entire fault accompanied by a dislocation on the surface shall be taken into account.* Then, it shall be to investigate in detail the adequacy of a fault postulated as the hypocenter, locational relationship with the site and the facilities established in it, and the adequacy of the determination of source characterization parameters. *While attention shall be paid to the applicability of the evaluation method based on the results of these reviews, the effect of various uncertainties on the evaluation of ground motions in ⑤ as above shall be evaluated in further detail and the standard seismic motion shall be formulated with a sufficient margin even based on the latest knowledge and findings related to the features of the earthquake ground motion extremely close to the hypocenter.*

g. The investigation and evaluation carried out when earthquakes for investigation are selected and the standard seismic motion is formulated shall be based on the latest scientific and technological knowledge and findings. With regards to existing materials, etc., their sufficiency and precision shall be fully taken into account and referred to. If the results are obtained that are different from existing evaluations, their bases shall be clarified.
h. If there is a facility with earthquake responses dominant in relatively long periods such as the use of the seismic isolation structure for the structure of the facility, attention shall be paid to its frequency characteristics to carry out the evaluation of ground motions and the standard seismic motion shall be formulated differently from that of other facilities.

(4) The "earthquake ground motion formulated without the hypocenter specified" shall be formulated according to the following policies.

a. When the "earthquake ground motion formulated without the hypocenter specified" is formulated, the propagation characteristics of a seismic wave up to the free surface of the basement stratum shall be reflected in the determination of a response spectrum when needed. The characteristics of the earthquake ground motion such as the duration of the earthquake ground motion, time history of the amplitude envelop, etc. shall properly be taken into account for the determined response spectrum.
b. The adequacy of the standard seismic motion formulated as the "earthquake ground motion formulated without the hypocenter specified" shall individually be confirmed against the latest knowledge and findings at the time of application. At this moment, with regards to the earthquake ground motion close to the hypocenter due to the causative fault without clear trace on the surface, the evaluation shall be referenced such as the probabilistic evaluation with various certainties taken into account.

(5) For the evaluation of ground motions based on the "earthquake ground motion formulated with the hypocenter specified for each site" and "earthquake ground motion formulated without the hypocenter specified", attention shall be paid to characteristic data required for an applicable evaluation method and the following items shall be taken into account related to the propagation characteristics of seismic waves.

a. In order to investigate the effect of the underground structure (deep and shallow ground structure) in and around the site on the propagation effects of seismic waves, the geological structure of such as the tilt of the layer, the fault and the fold structure shall be evaluated. In addition, the location and shape of a seismic bed-rock, the heterogeneity of a rock formation, and the underground structure such as the seismic velocity structure, and the damping characteristics of a soil shall be evaluated. Except the case where the underground structure is recognized to be made as stratification and homogeneous in the evaluation process, the three-dimensional underground structure shall be investigated.
b. Regarding with the investigation of the site and around required to implement the evaluation in (a), it shall be implemented to investigate the regional characteristics and existing literatures, to collect and analyze existing data, to analyze the observation records of earthquakes, to carry out geological, boring, and two-dimensional or three-dimensional geophysical surveys, etc. by combining them properly with appropriate procedures.

(6) Regarding with the "earthquake ground motion that is formulated with the hypocenter specified for each site" and the "earthquake ground motion that is formulated without the hypocenter specified", their corresponding probabilities of exceedance should be referred and it shall be grasped that their response spectrum corresponds to what level of exceedance probability.

(Reference)

The meaning of "to be referred" in detailed note on requirements (6) is intended to recognize the possibility of the occurrence of an event exceeding the earthquake ground motion that was formulated as described in (Reference) of detailed note on requirements of "1. Basic Policy Regarding Earthquake and Tsunami". It also means to pay attention properly to "residual risk" for the design of facilities and to continue make efforts to decrease it including the stages after the basic design stage.

<The following are major references.>

- Regulatory Guide for Reviewing Seismic Design of Nuclear Power Reactor Facilities (revision draft) (March 22, 2012)
- Guide for Seismic Safety Review of Nuclear Power Reactor Facilities (revision draft) (March 22, 2012)
- Preliminary version Reports on "Long Term Evaluation Method of Active Faults" (11/25/2010, Long Term Evaluation Subcommittee, Earthquake Investigation Committee, Earthquake Investigation and Research Promotion Office)

IV Seismic Design Policy

[Basic requirements]

Facilities shall meet the basic policy related to the seismic design by seismic design class as described as follows:

1. Each facility in Class S provided in paragraph 2 “Classification of importance of facilities” item 1 shall keep its safety functions for a seismic force due to the standard seismic motion. In addition, it shall endure, staying in almost elastic region, a larger seismic force either due to the elastic design earthquake ground motion or static seismic force.
   Facilities and SSCs in Class S provided in paragraph 2 “Classification of importance of facilities” item 2 and buildings and structures in which tsunami protection facilities are installed shall keep their functions required for each (tsunami protection, inundation protection and tsunami monitoring) against a seismic force due to the standard seismic motion.
2. Each facility in Class B shall sustain a static seismic force staying in almost elastic region. In addition, for those facilities that may cause resonance to occur, its effect shall be reviewed.
3. Each facility in Class C shall sustain a static seismic force staying in almost elastic region.
4. Safety function of an upper-level facility in each item as above should not be damaged by influence of lower level facility

[Detailed note on requirements]: omitted

V Considerations Regarding Ground Stability

【Basic Requirements】

1. Ground on which facilities are established shall have an adequate support capacity against expected seismic forces.
2. The safety functions of equipment and systems belonging to Class S in terms of importance in seismic design classification (hereinafter referred to as Class S equipment and systems) shall not be significantly affected by the tilt and folding of bearing ground (supporting soil) caused by crustal movement due to earthquakes.
3. Facilities having important safety functions shall not be seriously affected and lose its safety function by earthquake-induced abnormal conditions of surrounding ground, abnormal conditions such as subsidence uneven among buildings and structures (uneven settlement), liquefaction, consolidation settlement, etc.

【Detailed Note on Requirements】

- Regarding to the support capacity of the foundation grounds of buildings/structures that support Class S components and systems, *in addition to confirming that there is no outcrop of faults on which movements may occur in future as required by “1. Basic Policy Regarding Design against Earthquake and*

Tsunami", it shall be confirmed that there is no slip on any weak plane against the ground motions and that the supporting capacity against the standard ground motions is ensured.

- "Tilt and folding of bearing ground (supporting soil) caused by crustal movement due to earthquake" of Paragraph 2 include those caused by wide-area ground upheaval and subsidence as well as local ground upheaval and subsidence. Among these, *local phenomena of the tilt and folding of bearing ground may largely affect the safety and therefore particular consideration should be paid.*

<The following are major references.>

- Regulatory Guide for Reviewing Seismic Design of Nuclear Power Reactor Facilities (revision draft) (March 22, 2012)
- Guide for Seismic Safety Review of Nuclear Power Reactor Facilities (revision draft) (March 22, 2012)

VI Formulation of Design Basis Tsunami

【Basic Requirements】

Design basis tsunami used for facilities safety design shall be appropriate one based on the latest scientific and technological knowledge and taking into account the submarine topography from offshore tsunami source area to site vicinity and seismological viewpoint of geological structure and earthquake activities of offshore.

Design basis tsunami shall be determined while taking into account earthquake and other factors, such as landslide and slope failure, as well as their multiple combinations. Also uncertainty shall be taken into account and numerical analyses shall be carried out before determining design basis tsunami.

【Detailed Note on Requirements】 : omitted

VII Design Policy Against Tsunami

【Basic Requirements】

Since reactor facilities need to be highly safe against tsunami, they shall satisfy the following basic design policies so that they will not lose their safety functions even in the face of design basis tsunami:

1. A site where facilities provided in paragraph 2 "Classification of Importance of Facilities" item 1,having important safety functions, are located shall prevent design basis tsunami from directly reaching and intruding into them through

land area. Moreover, water intrusion into facility through intake and discharge channel etc. should be prevented.

2. Possibility of water leakage from any channel into water intake/discharge facilities, underground parts, etc. shall be considered. Ranges that may be flooded due to water leakage shall be limited to prevent such leakage from affecting the safety function of facilities.
3. Other than flood protection measures stated in subparagraph (2), flood protection measures for facilities provided in paragraph 2 "Classification of Importance of Facilities" item 1,,having important safety functions, shall be taken to isolate them from tsunami effect.
4. Effect of lowered water intake capability due to water level fluctuation on Facility safety functions shall be prevented.
5. With regard to tsunami preventing facilities and equipment provided in paragraph 2 "Classification of Importance of Facilities" item 2, their tsunami protection function and inundation prevention function shall be maintained against design basis input tsunami. And with regard to tsunami monitoring SSCs, their tsunami monitoring function shall be maintained against input tsunami.
6. Site upheaval and subsidence due to earthquakes, impacts of ground motions generated by main shocks and aftershocks, post-earthquake tsunami attack, repetitive tsunami attacks and tsunami-caused secondary effects (scouring, sediment transport, debris flow, etc.) shall be taken into account.

【Detailed Note on Requirements】: omitted

VIII Design Considerations Regarding the Stability of Surrounding Slope

【Basic Requirements】

A facility shall be designed so that its safety functions will not be seriously affected even when faced with the postulated slope failure surrounding it during earthquake.

【Detailed Note on Requirements】: omitted

Appendix B: Summary of "Regulatory Guide for Reviewing Seismic Design of Nuclear Power Reactor Facilities"

From: NSC Decision No. 2006-D59, Appendix 1, September 19, 2006, The Nuclear Safety Commission of Japan

(Disclaimer on the original document: This is an unofficial translation of the official Nuclear Safety Commission Document for the benefit of interested readers. For all questions regarding meaning and phrasing, please refer to the official version in Japanese.)

3 Basic Policy

Those Facilities designated as important from the seismic design standpoints shall be designed to bear the seismic forces exerted from the earthquake ground motion and to maintain their safety functions, which could be postulated appropriately to occur with a very low possibility in the service period of the Facilities from the seismological and earthquake engineering standpoints, considering such as geological features, geological structures, seismicity, etc. in the vicinity of the proposed site.

Any Facilities shall be designed to bear the design base seismic forces with sufficient margin, which are assumed appropriately for each seismic classification from the standpoint of radiological impacts to the environment due to earthquakes.

Buildings and structures shall be founded on the grounds which have sufficient supporting capacities.

5 Formulation of Design Basis Earthquake Ground Motion (DBEGM)

The ground motion to be established as the seismic design basis of the Facilities shall be formulated appropriately as the one, postulating to occur in a very low probability over the service period of the Facilities from the seismological and

Y. Suzuki, *Active Faults and Nuclear Regulation*,
https://doi.org/10.1007/978-981-15-0765-6

earthquake engineering point of view on geology, geological structures, seismicity, etc. in the vicinity of the proposed site, and having risks to give serious damages to the Facilities (the "Design Basis Earthquake Ground Motion (DBEGM) Ss").

DBEGM Ss shall be formulated on the following principles.

(1) DBEGM Ss shall be formulated as the following two types of earthquake ground motions in the horizontal and vertical directions on the free surface of the base stratum at the proposed site: The "Earthquake ground motions ((2) below) with the site specific earthquake source locations"; and the "Earthquake ground motions ((3) below) with no such specific source locations".
(2) The DBEGM Ss for the earthquake ground motions with the site specific epicenter shall be formulated on the following principles.

    (a) Earthquakes (more than one) are assumed which may have severe impacts to the proposed site, taking account of the characteristics of active faults, the earthquakes experienced in the past and at present in the vicinity, and classifying these earthquakes by their outbreak modes (hereinafter referred to as "Earthquakes for investigation").
    (b) Following consideration shall be made concerning the 'characteristics of the active faults around the proposed site' in (a) above.

        i) The active faults to be considered in the seismic design shall be identified as the one whose activities since the late Pleistocene epoch can not be denied. The faults can be identified depending upon whether or not the displacement and deformation exist by the faults in the stratum or on the geomorphic surface formed during the last interglacial period.
        ii) The active faults shall be thoroughly investigated by integrating geomorphological, geological and geophysical methods, etc. to make clear their locations, shapes, activity characteristics, etc. as a function of the distance from the proposed site.

    (c) For each "Earthquake for investigation" selected in (a) above, DBEGM Ss shall be formulated by the following two evaluation methodologies, respectively: i) with the response spectra; and ii) by the method with fault models. In evaluating the earthquake ground motions, sufficient consideration shall be made to the various characteristics due to the earthquake breakout modes, seismic wave propagation channels, etc. (including the regional peculiarities).

        i) Evaluation of earthquake ground motions with response spectra
        For respective "Earthquakes for investigation," response spectra shall be evaluated by appropriate methods and the design response spectra shall be defined based on these spectra. Earthquake ground motions shall be evaluated appropriately in considering their characteristics such as duration times, time dependent change of amplitude-enveloping curves suitably.

ii) Evaluation of earthquake ground motions by the method with fault models
For respective "Earthquakes for investigation," earthquake ground motions shall be evaluated by setting the epicenter characteristics parameters with appropriate methods.

(d) Uncertainties (dispersion) in formulating the DBEGM Ss as elaborated in (c) above shall be considered by appropriate methods.

Commentary: II Formulation of the DBEGM Ss

(1) Characteristics of the DBEGM Ss.
The Former Guide requested as the design basis earthquake ground motions two categories of Earthquake Ground Motion S1 and S2. The New Guide, however, these two Ground Motions were integrated as the DBEGM Ss, aiming at the enhancement of defining the Earthquakes for Investigation, evaluation of ground motions, etc.
This DBEGM Ss is the premise ground motion for the seismic design to ensure seismic safety of the Facilities. Its formulation requires sufficient checks of its adequacy, with reference to the latest knowledge in specific review cases.
(2) Relevant terminologies in formulating the DBEGM Ss

(a) The 'Free surface of the base stratum' for formulating the design basis earthquake ground motion DBEGM Ss is a hypothetically assumed free surface with no surface layers or structures thereon of the base stratum, which is almost flat with no significant unevenness and with a considerable expanse. The 'Base stratum' here is defined as a solid foundation, not significantly weathered, the shear wave velocity Vs of which exceeds 700 m/s.
(b) The 'Active faults' are those faults which have moved repeatedly in the recent geological age and may move in the future, too.

(3) Evaluation of the faults assumed as the epicenter

(a) An in-depth investigation should be made on the active faults, the basis of the evaluation of the faults to be assumed as the epicenter, combining the survey of existing reference materials, tectonic geomorphologic investigation, the earth's surface geological feature investigation, and geophysical investigation, etc. depending on the distance from the proposed site. Especially the area near the proposed site should be investigated precisely and in detail. The extent of the "area near the proposed site" should be defined suitably considering the correlation with the DBEGM Ss formulated as the 'Earthquake ground motions with no specific epicenter'.

(b) Active folds and active flexures, etc. relevant to seismic movement should also be included in the above-mentioned investigation (a) and should be considered in the evaluation of the faults assumed as the epicenter, depending on their properties.
(c) The properties of the faults should be evaluated appropriately by grasping the underground structures etc. in respective regions. Special consideration is required if the earthquakes are assumed from the properties of faults in the area where the faults are not distinctly locatable.
(d) If an empirical formula is used in assuming the magnitude of earthquakes from the fault lengths, etc., the unique features of the empirical formula should be duly considered.
(e) If the investigation of active faults is not powerful enough to collect sufficient information for setting the epicenter characteristics, including their shape evaluation, uncertainties incurred in setting the epicenter characteristics should be duly considered.

MIX
Papier aus verantwortungsvollen Quellen
Paper from responsible sources
FSC® C105338

If you have any concerns about our products,
you can contact us on
ProductSafety@springernature.com

In case Publisher is established outside the EU,
the EU authorized representative is:
Springer Nature Customer Service Center GmbH
Europaplatz 3, 69115 Heidelberg, Germany

Printed by Libri Plureos GmbH
in Hamburg, Germany